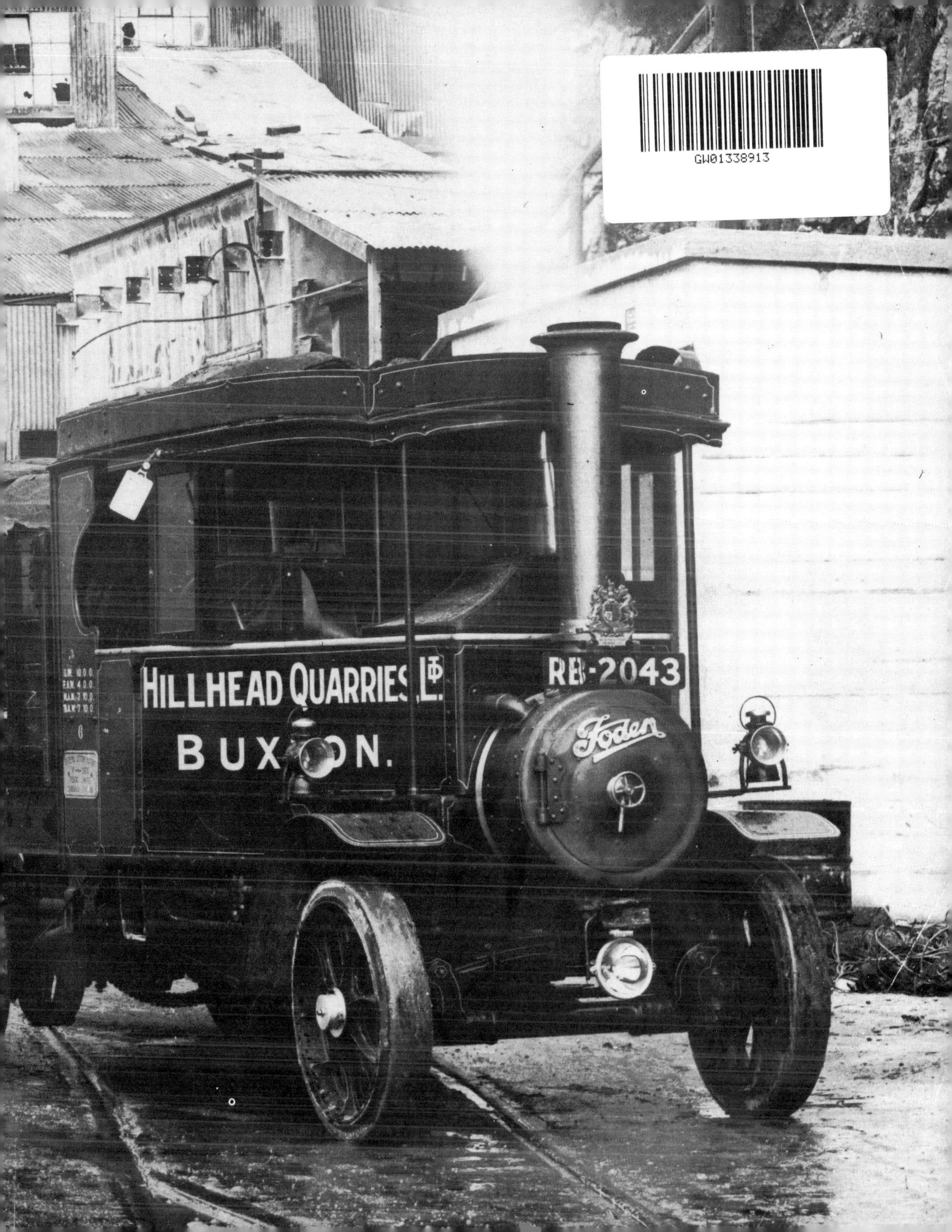
HILLHEAD QUARRIES LTD
BUX ON.
RB-2043
Foden

# THE Focus STORY

**PSL** Patrick Stephens, Cambridge

*From farm machinery to diesel trucks*

# THE Foden STORY

*Pat Kennett*

© Pat Kennett 1978

All rights reserved. No part of this publication may be reproduced, stored in a retrieval system or transmitted, in any form or by any means, electronic, mechanical, photocopying, recording or otherwise, without prior permission in writing from Patrick Stephens Limited.

First published in 1978

**Front endpaper** *Hillhead Quarries at Buxton was a staunch Foden customer for many years. This is a 12-tonner, equipped with the optional enclosed cab, hauling out of the quarry in about 1928. A Sentinel steamer can be seen in the loading bay while on the left is a Manchester tipper — the derivative of the Willys-Overland-Crossley (see Chapter 6).*

**Back endpaper** *The eight-wheeler Foden was a consistent market leader in its weight class from the inception of the 30-ton rigid-eight truck.*

**British Library Cataloguing in Publication Data**

Kennett, Pat
    The Foden story.
    1. Fodens Ltd — History
    I. Title
    338.7'62'92240942713    HD9710.G74F/

ISBN 0 85059 300 X

Design by **Tim McPhee**

Text photoset in 10 on 11pt Times Roman by Blackfriars Press Limited, Leicester. Printed in Great Britain (on 100 gsm Fineblade coated cartridge) and bound by The Garden City Press, Letchworth, for the publishers, Patrick Stephens Limited, Bar Hill, Cambridge, CB3 8EL, England.

# Contents

Preface 7
Chapter 1 Sowing the seed 9
Chapter 2 The birth of the Foden Age 14
Chapter 3 The great industrial engines 22
Chapter 4 Working with the Victorians 32
Chapter 5 Wagons and the War Office Trials 37
Chapter 6 Into the golden era 47
Chapter 7 A symphony in steam 58
Chapter 8 Men of steam 69
Chapter 9 Steam down below 78
Chapter 10 Not so much a job, more a way of life 87
Chapter 11 The Foden Motor Works Band 99
Chapter 12 The spawning of ERF 109
Chapter 13 A painful transition 115
Chapter 14 The return of Mr William 126
Chapter 15 War and peace 133
Chapter 16 The postwar expansion 146
Chapter 17 Expansion, crisis, recovery 161
Chapter 18 Into a new age 173
Appendix Principal members of the Foden family 179
Index 181

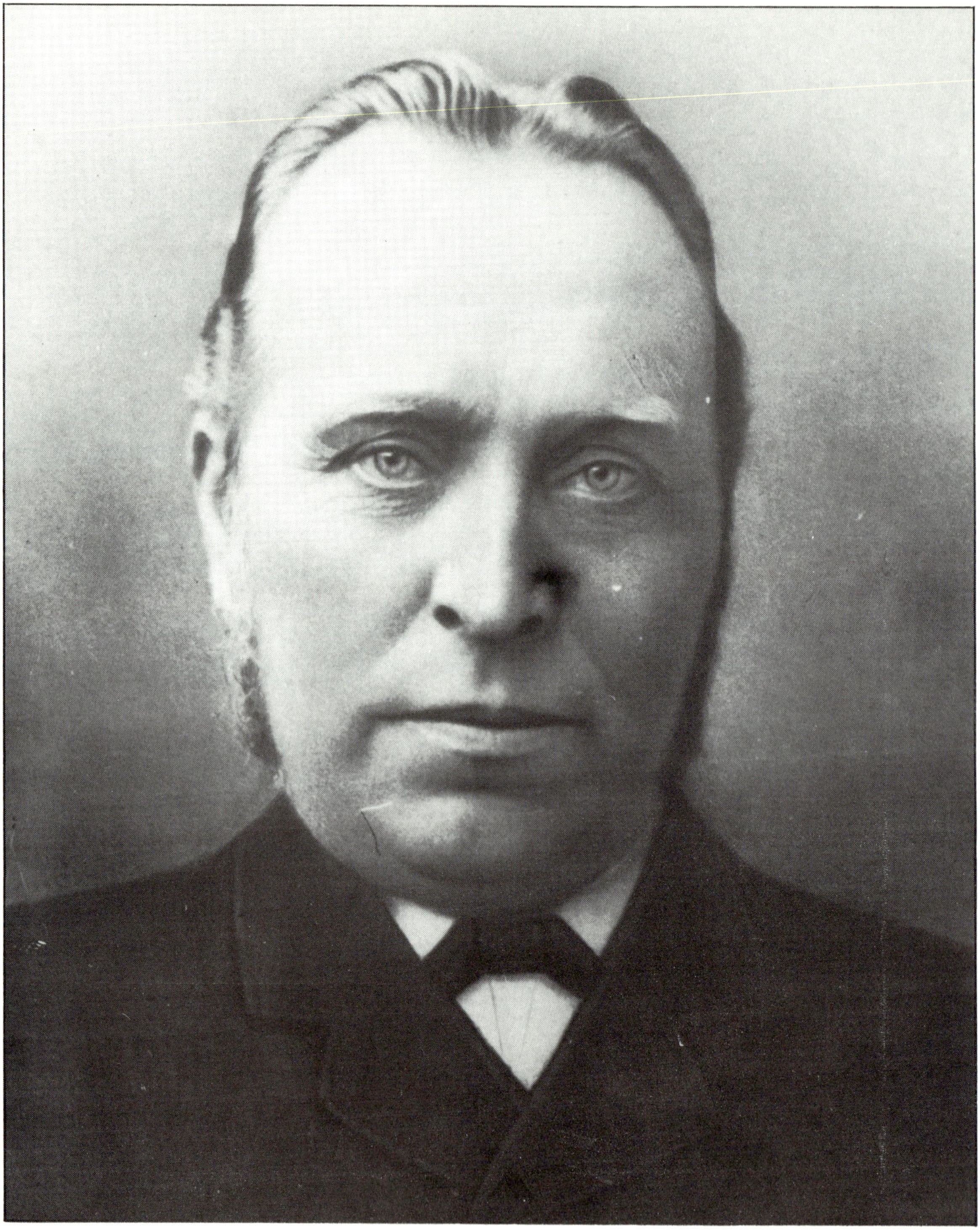

# Contents

Preface 7
Chapter 1 Sowing the seed 9
Chapter 2 The birth of the Foden Age 14
Chapter 3 The great industrial engines 22
Chapter 4 Working with the Victorians 32
Chapter 5 Wagons and the War Office Trials 37
Chapter 6 Into the golden era 47
Chapter 7 A symphony in steam 58
Chapter 8 Men of steam 69
Chapter 9 Steam down below 78
Chapter 10 Not so much a job, more a way of life 87
Chapter 11 The Foden Motor Works Band 99
Chapter 12 The spawning of ERF 109
Chapter 13 A painful transition 115
Chapter 14 The return of Mr William 126
Chapter 15 War and peace 133
Chapter 16 The postwar expansion 146
Chapter 17 Expansion, crisis, recovery 161
Chapter 18 Into a new age 173
Appendix Principal members of the Foden family 179
Index 181

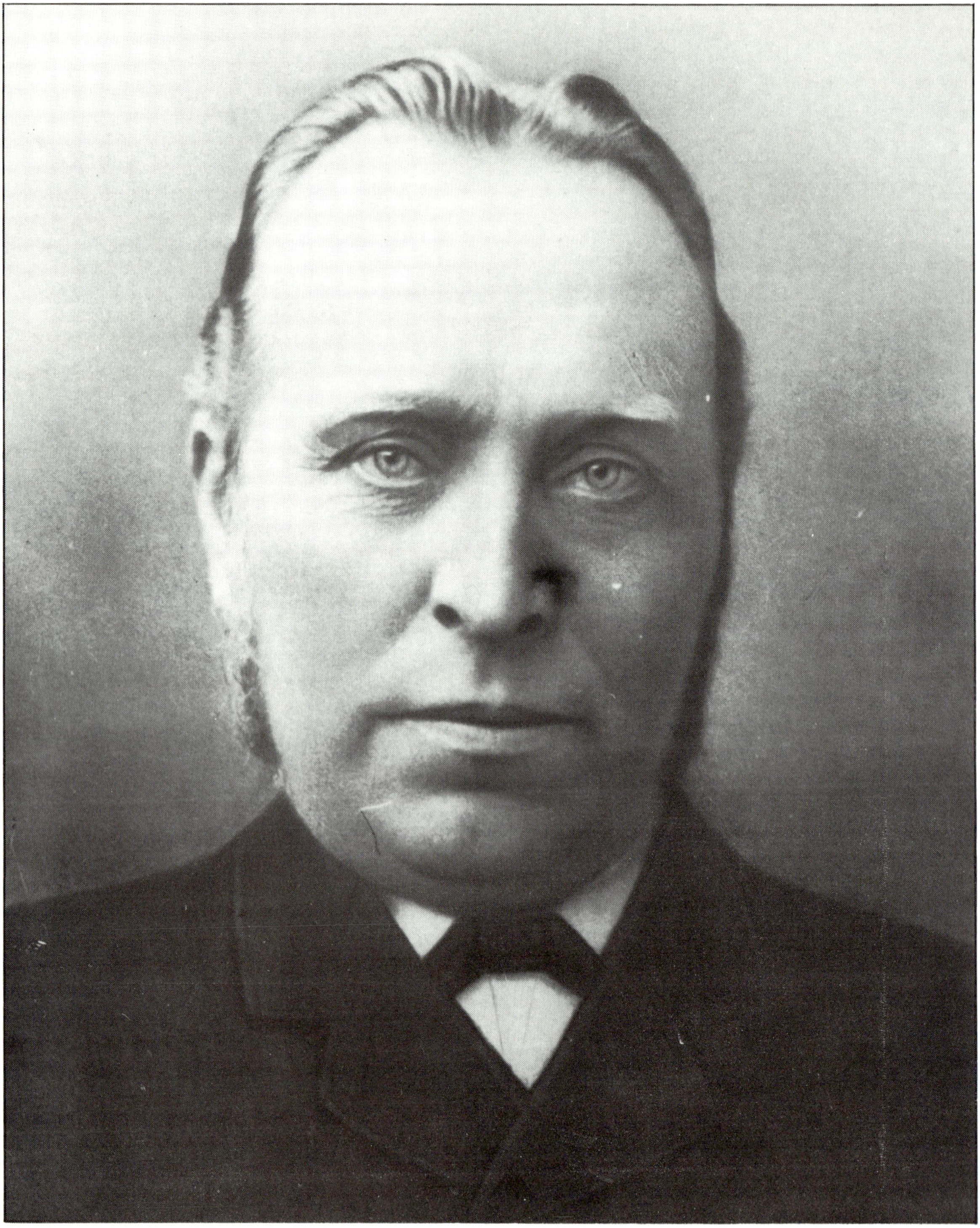

# Preface

In researching a book like *The Foden Story,* it is essential to 'live' the subject in order to portray the correct shades of feeling and meaning, especially in those sections dealing with the social rather than engineering aspects of the story. The 'living' of that story was a fascinating experience and it was almost with a sense of regret that I had to wake myself up and return to this modern age, once my work was completed. There can be no doubt that the latter part of the 19th century and the early decades of this one were momentous times for Britain, indeed much of the civilised world, as man developed his skills to higher levels than ever before, yet found himself struggling ever harder to avoid the miseries of poverty and deprivation. *The Foden Story* tells how one family and its brilliantly successful products contributed to those struggles and triumphs, for more than a century. In relating that tale from the very beginning, I have been helped by all manner of people from all walks of life.

In the early stages of the investigations into the origins of the company's activity, I was deeply indebted to the late Mr Tom Hancock of Perth, great-grandson of Walter Hancock, who helped establish the link between Hancock's steam carriages of the 1830s and the subsequent Foden company. Then I was fortunate in tracking down an eminent local historian in south-west Wales, W. H. Morris, who proved to be an enthusiastic and prolific source of facts concerning the social and industrial life of the area, including the Foden engines at Kidwelly. From his clues I was able to establish that there was a great affinity between the Chivers and Foden families.

Numerous Foden employees, past and present, were keen to help me set out the detailed progress of the firm over the years, and it is perhaps invidious to pick out individuals. However, the contributions that stand out among the hundreds of sheets of notes, and seemingly miles of recording tape that were necessary to sift all that information into some kind of order, came from all levels. From the management end of the spectrum, Ted Foden and Ted Johnson spent a great deal of time recording their reminiscences, while Clifford Brassington, who was chief buyer for many years, recalled all manner of problems and escapades in financial matters. Many band stories came from Don Burgess, who has himself written a book about the Foden Band. Ted Gibson and Harold Rancollis filled in a lot of the technical detail and stages in development of the products. Harry Smith remembered what it was like as an apprentice in the early days of making steam wagons, and retold all that in graphic style. He also produced many interesting artifacts from those bygone days. Albert Bourne recalled the skills and endurance needed to be a successful steam wagon operator, and demonstrated that he still has those skills, despite being long retired.

Throughout, the Foden management gave me full access to records and files, which revealed a considerable degree of trust in view of the fact that I earn my living primarily as a journalist. To them I have to say a special thank you for making the

**Left** *Edwin Foden, a great Victorian and a great engineer, was born in 1841, son of the village shoemaker at Smallwood in Cheshire.*

task possible in the first place. While all that was going on the seemingly tireless Aubrey Evans, who manages the publicity office at the Elworth works, never hesitated to offer help, leads, suggestions or material when innumerable new lines of enquiry were mooted over a period of about 18 months. I am sure his office must be a lot more productive now that I have finished my book! To all those other friends and colleagues, both inside and outside the company, who contributed to the work, not least the long-suffering Carol who typed the manuscript from my drafts, amid the chaos that is my office, my very sincere thanks.

Of course, even in a book of this size, it has been quite impossible to enter deeply into the engineering and commercial detail that many readers find so absorbing. Even were we dealing with just one period, say the steam wagons, the many variations among the more than 7,000 built would defy the limitations of a single volume. The same can be said for the diesel age from 1931 onwards, when there was such prolific development. Only the principal steps in that process can be outlined here, and there simply is not the opportunity to delve into the details, fascinating though they may be. What *The Foden Story* sets out to do is to present an overall picture from the days when the firm's activities were still the far-off dreams of a small boy, up to the harsh realities of the present industrial world. That picture not only involves the products and the engineering, but people are a vital part of the scene too. Their living standards, their recreation, their habits make a story that is more vital than mere machinery could ever be. The relationships between employer and employee in a close community cannot be ignored, neither can the activities of one of the most famous musical ensembles of all time, the Foden Motor Works Band. All those facets of life around Elworth are as much a part of *The Foden Story* as the War Office Trials, or the financial crisis of 1974.

I trust that my readers will gain some measure of enjoyment from this book, perhaps even a few crumbs of education too. Certainly there was a great deal of enjoyment involved in putting all those words together, tracking down suitable pictures and taking new photographs. At the time I felt it was rather hard work, but I changed my mind when it became clear just how hard those Victorians and Edwardians worked. I trust that my humble words may be considered to do justice to their labours.

## Chapter 1

# Sowing the Seed

Edwin Foden was undoubtedly one of the great Victorian engineers. That his contribution to the golden age of the machine came rather later than some, does not diminish its impact or importance. Along with Stephenson, Bessemer, Whitworth, Brunel, Lanchester, Parsons and many others, Foden carved his immortal niche in the history of an era during which Britain rose to world supremacy in engineering matters both large and small. Edwin Foden was a son of the age of steam, and to him steam was the source of all things good in the world, be it power to drive manufacturing or processing machinery, or a means of transporting those products to their best marketplace. His massive contribution to industrial and commercial strength has origins which can be traced back much earlier than the turn of the 20th century, when Foden became a household name, indeed back to the first half of the 19th century, before Edwin was even born.

In the late 1820s and early '30s, before Britain's railways developed into a great national transport system, one Walter Hancock designed, built and patented a number of steam carriages. Quaint though those early machines may now appear, they were ingenious and immensely successful. Almost a century before the charabanc outing to Brighton was considered a popular pastime, Hancock and his colleagues who operated those carriages from London were running excursions to Brighton, as well as shorter trips to Hampstead Heath and the Downs in Sussex and Surrey.

Walter Hancock had his workshops at Stratford — then a small town some miles east of London — and his first successful carriage, named *Infant,* was built in 1826. *Infant* became the first steam road vehicle to carry passengers for hire, the first to make the journey to Brighton, and the first to carry passengers in the City of London. Other carriages followed, such as *Autopsy* which was built in 1833 as a steam omnibus to run regular services between Pentonville, the City and Paddington. Perhaps the most impressive of all the Hancock carriages was *Enterprise,* a big steam omnibus capable of carrying 20 or more passengers. It was built in 1833, and ran for several years on various routes in London, notably Paddington to the City. Hancock's workshops built 11 successful large steam vehicles in all, and numerous small ones, as well as all manner of ingenious engineering devices.

What, you may ask, has all this to do with Edwin Foden, the son of a Cheshire shoemaker, if it all happened before he was born in 1841? Simply this. After the steam carriage was forced out of the picture by the railways — particularly after Queen Victoria gave royal support to the railways by making an ostentatious and much publicised journey from Slough to Paddington on June 13 1842 — Walter Hancock retired, and his son George, who had worked on the great carriages with him, was obliged to seek his own fortune elsewhere. He went far from London, and eventually set up an agricultural engineering workshop with a friend in rural Cheshire. That was in 1848, and the company was called Plant & Hancock. In 1856 Edwin Foden, now 15 years old, became an indentured apprentice to that firm. The combination of George Hancock's experience of many years in the practical aspects of motive steam, and the brilliant engineering brain of Edwin Foden, made an impressive team. Edwin widened his experience in Crewe railway workshops and in another workshop at nearby Kidsgrove before returning to Elworth, and by the time he was 19 he was shop foreman. By 1866 Hancock had made young Edwin a partner in the firm, which was then renamed Hancock & Foden, although everyone referred to their works as the Elworth Foundry.

Just why George Hancock should have decided to go to Cheshire to establish his business is not recorded. But undoubtedly he saw that agriculture was the best outlet for his expertise in engineering and steam, given that he did not have the resources to compete with the great railwaymen like Brunel and Stephenson, and that successive acts of legislation made road-vehicle prospects extremely poor in comparison with those of the railways.

In the eastern and southern farming regions of England, other agricultural engineers were already gaining strength. Charles Burrell & Sons were at Thetford in Norfolk, Richard Garrett was at Leiston in Suffolk, Ransomes were established at Ipswich, Clayton at Lincoln, William Allchin at Northampton, and John Fowler at Leeds. Further south, Aveling appeared in Rochester, and William Tasker in Andover, while Wallis & Steevens were close by at Basingstoke. To the west there were only Frederick Danks at Oldbury and Naylers at Hereford. Most of these were established before the Hancock steam carriages and omnibuses finally disappeared, and George,

*An Ackerman engraving of 1833 shows the London & Paddington Steam Carriage Co's* Enterprise *at full speed. The livery included the legend 'Walter Hancock Patentee'.* Enterprise *entered public service on April 22 1833. The boiler was at the rear with an undertype twin-cylinder engine driving the rear wheels.*

*Sowing the seed*

who was always a very logical thinker, apparently chose the one large remaining area of agricultural countryside without a major steam engineer. Fate completed her work by bringing him into contact with Edwin Foden, and changed the course of engineering for many years to come.

George Hancock's original small engineering workshop was established in what was then a mainly rural area, about 200 yards from the village railway station at Elworth. This village was near the old market-town of Sandbach, and a few miles north of a new railway-town then beginning to grow, called Crewe. It had occurred to Hancock that to build an engineering establishment out of all contact with other engineering activity might prove to be too much of a problem when it came to labour recruitment. But Crewe provided a potential pool of skilled and semi-skilled labour not only to Hancock but also to a number of other small engineering concerns in the Cheshire and Staffordshire area. The railway also represented a means of transport for both the necessary raw materials and the finished products. It is indicative of the man that Hancock was not too stubborn to make full use of the system that had earlier forced him and his father out of business.

When Edwin Foden, as an enthusiastic youth of 15, came to work at Elworth Foundry — walking the three miles to and from his home at Smallwood night and morning — the works was making a great variety of iron and steel products. Agricultural power units known as 'portable engines' were just one of the lines, along with tilling machinery, threshers, chaff cutters,

*An 1832 lithograph shows Walter Hancock's steam carriages* Autopsy, Era, *and* Infant, *three of the 11 he is known to have built. The vehicle on the hill in the background is believed to be the 'drag' built for a Dr Voigtlander of Vienna.*

rootcrop pulpers, pumps, and all the other impedimenta of Victorian farms. The firm later made a huge variety of industrial steam engines for mines, mills, canal pump-stations, water works, sewage plants, indeed for almost any kind of manufactory that needed power, and these were called fixed engines. When such jobs were scarce, they filled in with sub-contracting work for the railways down at Crewe, and built an occasional ship or machine tool. When work was hard to come by they even sank wells for the Cheshire brine-salt industry.

There were undoubtedly long and nostalgic discussions between George and Edwin for, despite their considerable age difference, the two men got on very well together, a fact which is amply illustrated by Edwin's rapid promotion in the firm and his establishment as a full-scale partner only ten years after joining as a new apprentice. George Hancock was realistic enough not to hanker after those spectacular steam carriages of years gone by, dear to his heart though they were. They were gone for ever. For its passenger transport, the public had overwhelmingly chosen the railway. If the railway was good enough for their beloved Queen Victoria, it was good enough for them, they reasoned, and numerous prominent politicians spoke out along similar lines, to some extent jumping on a vote-catching bandwagon.

The roads were generally in a very bad condition anyway, since few were the responsibility of the local authorities, being mainly left to the whims of the owners over whose land they ran. Even the main turnpikes which once formed the mainstay of internal communication had long since been left to deteriorate over much of their length, and the good roads were almost entirely confined to towns and cities, and to certain areas where road builders like John Metcalf and John McAdam had been asked by enlightened local authorities to improve matters. In addition, the pro-railway fervour that had swept the country led to punitive legislation against any form of mechanically propelled road vehicle. The railways were the place for machines, successive politicians argued, and roads should be preserved for the use of horses.

Consequently, it came as little surprise when, in 1861 — Edwin Foden was in his fifth year with the firm at Elworth — the Locomotives Act was passed by Parliament. This act made it mandatory for any mechanically propelled vehicle used on a public road to be preceded by a man carrying a red warning flag. Furthermore it imposed speed limits of 4 mph in the country and 2 mph in towns, and this clearly ruled out the economic use of such a vehicle for anything other than a very short journey.

However, a number of steam-powered agricultural locomotives and tractors were built by various engineers in Britain, including Hancock, but their efforts were immediately squashed by the passing of an even more discriminatory and restrictive act, the Locomotives Act of 1865. This new act made it illegal to operate any heavy self-propelled vehicle, whether for agriculture or any other purpose, on a turnpike road outside the hours of midnight to 6 am, and made road machines the subject of all manner of bridge restrictions, with heavy fines for infringements. Such restrictions were intolerable to the owners of these machines, and constituted an open charter for any local authority or other body — including railway companies — wanting to restrict road traffic. Consequently, there was no incentive to use road locomotives or wagons, let alone to design and build new ones. In contrast, the rapid expansion of the railway network was accompanied by several encouraging Acts of Parliament which made the purchase of privately held land and property, and the provision of masonry and other materials from local resources, into an almost divine right of the railway boards, whose progress was now thought to be of paramount importance to the nation.

It was against this background that Edwin Foden matured his engineering training and developed his ideas. Clearly the use of his beloved steam power for motive purposes was out of the question under the current law, but he remained convinced that one day the era of the steam road vehicle would come. So much promise had been shown by the Hancock machines of 30 years earlier, that both partners were certain that mechanical efficiency and convenience in an increasingly industrialised nation would eventually overcome the prejudice that had set the legislative pattern of the 1860s and '70s. But a further blow was yet to come. In 1867

*Sowing the seed*

amendments to the Locomotives Acts prohibited the use of an agricultural engine, whether stationary, portable or powered, within 25 yards of a public roadway, and demanded the presence of a man in the road to signal the operator to stop the engine whenever a horse approached.

Under such illogical and irksome conditions, it seems miraculous that the agricultural mechanisation revolution ever got under way at all. But such was the ingenuity and dedication of men like Foden and his colleagues in all the famous agricultural engineering centres — Tasker, Fowler, Garrett, Ransome, Allchin, Clayton and the rest — that not only did the industry survive, but for many years it kept pace with the general manufacturing industry in Britain, ensuring that agriculture was still a developing and expanding activity. The days of the small farmer with his horse and scythe were gone. A sizeable demand for food and produce at competitive prices meant a high degree of mechanisation, and efficient mechanisation at that. In later years that mid-Victorian trend was to be slowed down, even reversed in some areas, but in the last three decades of the 19th century mechanisation was the pattern of industry.

And it was the pattern of the economic outlook when George Hancock retired from the partnership in 1870, remaining as a consultant for a short time. The firm had been renamed Foden & Hancock, and remained so for some years after George Hancock had ceased to take any active part. But in 1868, less than two years after Edwin had won his partnership in the firm, he was presented with a son, and in anticipation of young William joining the firm as soon as he was old enough, in 1876 Edwin renamed the company Edwin Foden & Son, although it was to be some years, obviously, before William could take any part in the proceedings. In the meantime, Edwin traded variously under the name of E. Foden, Edwin Foden, Elworth Foundry, and Edwin Foden at Elworth Foundry. Under those titles, engineering products of great variety and remarkable efficiency were made and sold in every part of Britain. They had one thing in common — they were all exceptionally well made. Edwin Foden, born of craftsman and brought up in a Methodist family environment, was a firm believer in the notion that if a job was worth doing, it was worth doing well.

## Chapter 2

# The birth of the Foden Age

Left to run the company and form his own ideas, without first having to explain them to his partner, Edwin Foden thrived. Still not yet 30 years old, with all the vigour of youth, the stimulus of his natural inventive wit and ingenuity, and the experience of an old hand like Hancock committed to memory, he set about improving the efficiency of his products, not content with the adage that the old ways are always the best. So, while the works got itself busy with proven designs that were in steady demand as a result of the firm's already sound reputation, Edwin Foden spent more and more time experimenting and researching with new designs and new engineering concepts.

*Foden threshing machines formed a significant part of the total output in the 1870-80s, some powered by a 'portable' engine, others by their own small steam engine.*

The process by which steam was admitted to a single cylinder to produce power, then exhausted and for all practical purposes wasted, worried Edwin. It took precious, expensive fuel to raise that wasted steam, and the economics of farming and industry were scarcely less critical then than they are now. His answer was a compound engine, which took six years to develop to a commercial reality from 1880, when he patented its design for use in locomotives. His engine used the exhaust steam from a high-pressure cylinder to fill a larger and lower-pressure cylinder, thus utilising much of the energy that would otherwise have gone to waste. In fact, the power output of these compounds was about 70 per cent greater than that of the equivalent single-cylinder engine from a similar weight of fuel. He had developed compound engines in fixed applications some years before, but his locomotive compounds were very advanced.

Throughout those early years on his own, Foden concentrated on power plants for agricultural and industrial use. The agricultural machines were initially what were called 'portable' engines. They had a boiler and a chimney, much as the later traction engines and locomotives did, and were fitted on wheels for mobility. The engine was mounted along the top of the boiler, and supplied the power through a belt driven from the engine's flywheel. Why then were horses needed to move these 'portables' around? The reason lay in the legislation that was heaped upon the road vehicle throughout that period. It made more sense to use the portables as horse-drawn vehicles because, even though they developed considerable

## The birth of the Foden Age

steam power, they could not then be called locomotives or steam tractors, and were classified simply as horse-drawn machinery. This exempted them from much of the petty restriction on the time of day when they could be moved on the road, most of the bridge restrictions, and even, in some instances, from the speed restrictions; and, of course, the red-flag man was not needed either.

The portables were hauled to wherever they were needed, and then used for ploughing, harrowing, draining, threshing, pumping, and other heavy work around the farm. The various means by which they worked were ingenious. Ploughing, for example, was achieved by driving a winch from the engine, which hauled a plough backwards and forwards across the land with the aid of cables and pulley tackle. In fact, John Fowler of Leeds perfected and refined the principle of the steam ploughing engine more than anyone else, and his pairs of gigantic ploughing engines built from around 1880 onwards — some surviving to this day — were perhaps the most spectacular products of the steam age in agriculture. But they were expensive machines, and the more modest devices which Foden produced were within the reach of many more farmers, and were consequently built in larger numbers. His agricultural engines began as single-cylinder machines, and later employed the twin-cylinder compound engine, acquiring an excellent reputation for reliability as the years went by.

*The first big Foden engines were built in the early 1880s and had single-cylinder engines. They were designed for agricultural use, since legislation virtually prohibited road haulage of freight. This picture is reproduced from an original wood cut.*

**NEW PATENT TANDEM COMPOUND TRACTION ENGINE.**

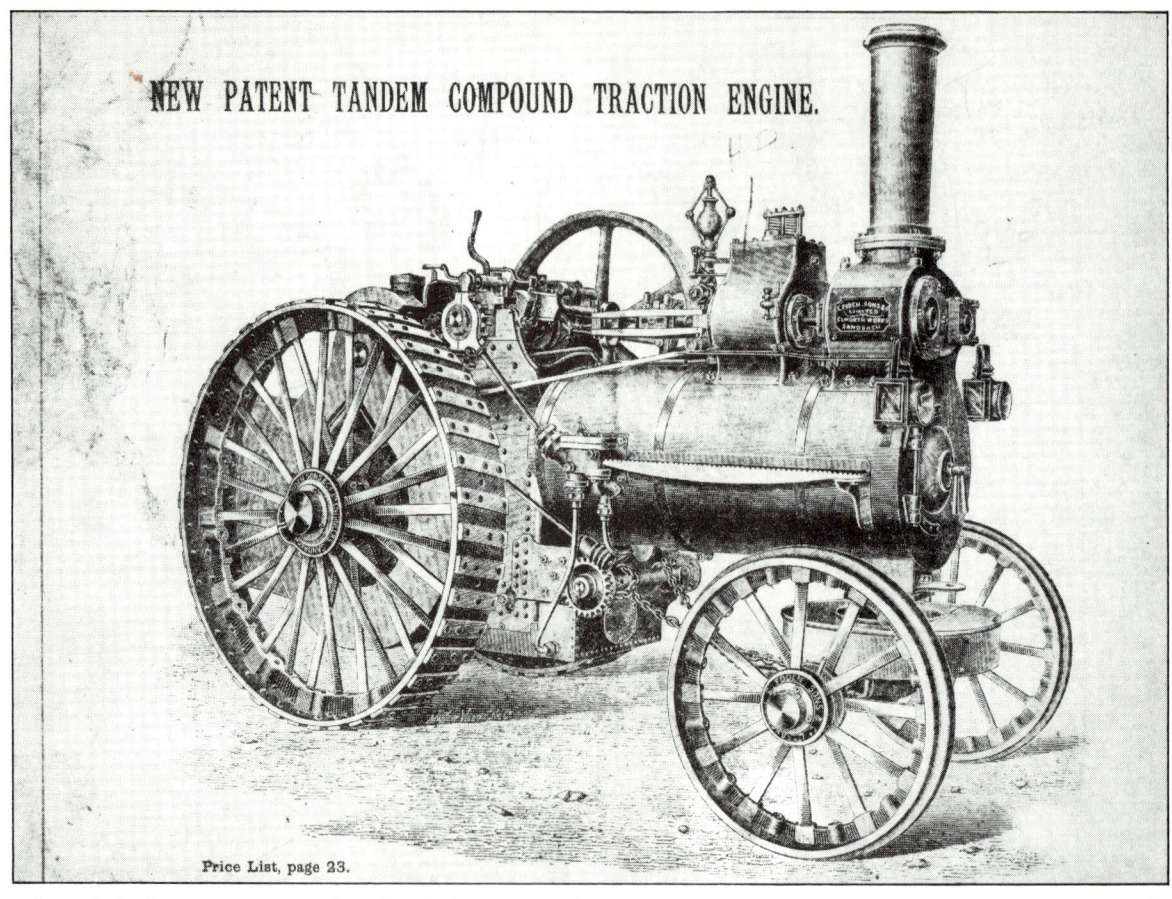

Price List, page 23.

*Edwin Foden's clever compound-engine design appeared on a traction engine in 1887. More elaborate versions for agricultural use had a special control on the valve gear to enable the engine to start under load, as distinct from hand-rocking on the flywheel.*

Many farmers either could not afford or could not be bothered with all the winches and tackle that went with mechanised arable farming, and so they stuck to the tried and trusted horse team for that work. However, the mechanical threshing machine was a much more attractive proposition. The changeable British weather meant that harvesting and threshing the grain crop was a time-sensitive operation. The old hand-methods with horse- or donkey-powered systems were not fast enough to get the crop under roof and dry within any average spell of good weather. The answer was the mechanical threshing machine. Many of those used in mid-Victorian Britain were driven by belts from a borrowed 'portable'; indeed machines of a very similar type driven by steam traction-engines — and later internal-combustion-engined tractors — were in use until quite recently when the combine harvester eventually took over.

Edwin Foden decided that there was a market for mechanical threshing machines, even if the farmer did not have or did not want to have a portable engine. His solution was to build a drum thresher powered by its own small steam engine and usually horse-drawn. It proved a resounding success. Foden used the principle of boosting the air blast through the boiler tubes by exhausting the cylinder up the smokestack, a common enough idea in later years, but virtually unknown in small engines at that time. These machines escaped much of the anti-engine legislation and did a fine

*The birth of the Foden Age*

*Foden anticipated some relaxation in the road legislation in the 1890s, and added road locomotives to the catalogue. Basically they were very similar to the single-cylinder agricultural engines, but with faster gearing and a canopy.*

job of threshing grain of different types.

Fire precautions must have been important, because there are few places as inflammable as the threshing yard. The cut sheaves, the flying chaff, the discarded straw, to say nothing of the sacks of grain and the wooden casing of the machine itself, were all capable of catching fire spectacularly. But there are no records of any serious disasters of that kind, and Edwin Foden's threshing machines went on to win a gold medal in 1883, at the Royal Agricultural Society's Show at York that year. In all, about 50 of them were made between 1882 and 1891. The price ranged from £140 up to about £180, depending on the size.

The influence of agriculture at that time was considerable, and the voices of both the wealthy landowners and the less wealthy farmers began to increase to a pitch that politicians could ignore at their peril. A total of 15 acts of Parliament were passed between 1860 and 1895, all adding in some way to the restrictions on the use of engines on roads, and mostly including agricultural engines. The farmers were getting thoroughly tired of legislation being continually levelled against them, and the economics of their entire industry were now at stake. Sensing the rising tide of rebellion on the land, Mr Benjamin Disraeli's government rushed through an act in 1878 which did away with the need for a man with a red flag to precede any agricultural engine on the road. It also removed the irksome restrictions on the number of hours that an agricultural engine could be used in a day.

These restrictions were not abolished for non-agricultural engines, so there was still a virtually total ban on anything resembling an industrial commercial vehicle on the roads.

The added scope that the agricultural engine concessions gave to designers was not overlooked by Edwin Foden. He had already been experimenting with means of improving the efficiency of his stationary engines. There was now clearly a good market for traction engines as well, so fuel economy was more vital than ever, because all the fuel would have to be carried on the engine instead of being left in a heap at the work site. His twin-cylinder compound engine was the first major step, and it took him three years to develop his patented design into a functioning traction engine. In fact it did not initially work as well as it might have done, and consequently many of the earlier traction engines sold to farmers and agricultural contractors from 1882 onwards were single-cylinder types.

It was not until 1887 that a reliable, marketable compound engine was produced, and Edwin was so pleased with it that he entered it in the Royal Agricultural Society Trials at Newcastle-upon-Tyne that year, where he competed with the major agricultural engine manufacturers from the great farming flat lands of eastern England. The Foden engine won a gold medal by virtue of its low fuel consumption and flexibility in performance. In fact its consumption was so far below the norm for such machines (at 1.8 lbs of coal per horsepower per hour) that several authorities suggested that Edwin and William —

Pride of Leven *was one of the earliest compound engines built, and still runs admirably despite being nearly 100 years old!*

*The birth of the Foden Age*

*When road vehicles were virtually banned by pressure of legislation, agricultural work kept the company alive. Among their products was a high-capacity straw baler.*

who by then was accompanying his father on such expeditions — had somehow cheated. But there was no cheating, and that efficiency, coupled with the fact that the impressive engine cost only £165 — well below average at that time — made the Foden engine an attractive machine. The secret lay in design and construction simplicity, arising from Edwin Foden's genius in refusing to use two parts if their separate tasks could be performed by one. That is undoubtedly the reason for the considerable gestation period of six years between the original patent and the final saleable machine. But it was worth waiting for.

Despite the very competitive situation in the industry, Edwin was delighted to find that orders for his new machine came pouring in. New manufacturing plant had to be installed, and steps had to be taken to speed up production without in any way compromising the high engineering standards that made the engines so successful. Most engines were built for farm work, one or two for haulage purposes and an occasional example for travelling showmen. One of these — *Prospector* — is preserved to this day. But Edwin was reluctant to rely in any way on agricultural engines. He therefore retained the heavy industrial engine side, and continued to experiment with ideas that would eventually lead to proper road freight vehicles, once new legislation opened up the market.

Even though the 1878 agricultural concessions were welcome, they were granted in an atmosphere which suggested that they would be withdrawn at the slightest sign of misbehaviour on the part of those users who had the temerity to operate their infernal steam contraptions on the highways. Indeed, one of Fodens' competitors, William Tasker, wrote in a handbook published for the guidance of agricultural engine users '... although some official restrictions have been removed, and much public prejudice overcome, very great care and judgement must be exercised in working these labour saving machines. Unfortunately many local and other authorities still regard steam traffic upon roads as a nuisance to be put down if possible, or to be impeded where extinction is not possible... the fire should not be fed when passing, or just before passing, through towns or villages... uncivil or ill-tempered men should on no account be entrusted with the management of road locomotives...'.

Many readers may be forgiven for thinking that the so-called anti-'juggernaut' campaign is a

*Sister-engine to* Pride of Leven *seen working on a farm at the turn of the century. These compound engines were both powerful and economical.*

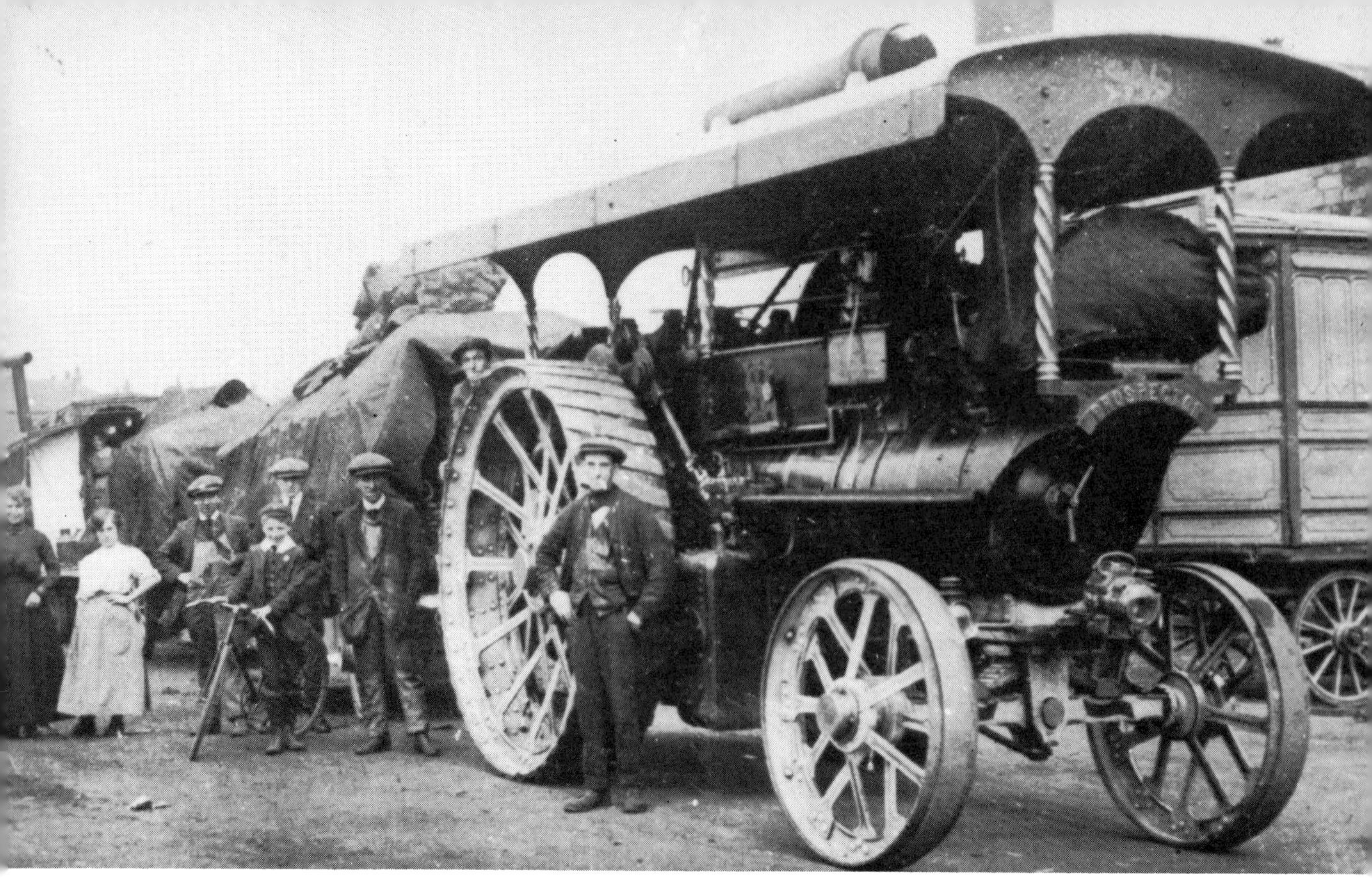

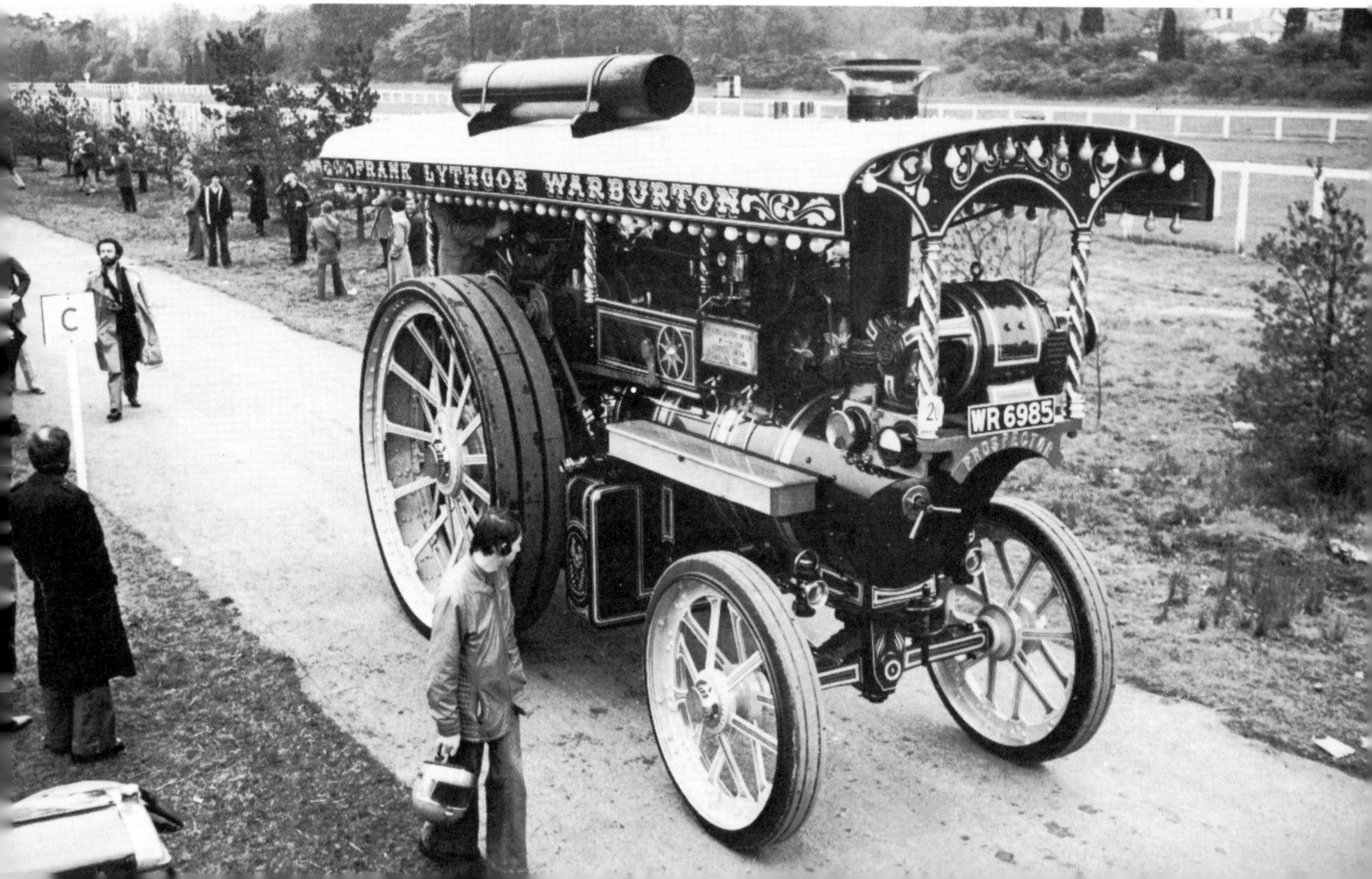
FRANK LYTHGOE WARBURTON
C
WR 6985
PROSPECTOR

modern phenomenon, dreamed up by the press of the 1970s to provide ammunition for readers requiring a popular 'Aunt Sally' on which to expend their energies. But the habit is at least a century old. In 1878 several long articles in *The Times* spoke of the 'tyranny of the roads', and complained about the dangers to property from road vehicles by vibration and fire, and of the fact that they frightened animals. This was directed at all road users; *The Times* even fulminated against the dangerous and excessive speeds of cyclists!

*The Daily Telegraph* appears to have been the first to apply the term 'juggernaut' to the unloved road vehicle, in an article published in 1897. Among other things the article said, '. . . it is time to put these uncharted libertines of locomotion in their proper place. They drive their machines as the ironclads of the highway, intent on ramming the enemy, every day becoming more arrogant, more reckless, more indifferent to everyone's inconvenience but their own. The danger to personal security has become a genuine one, the question of damage to public property and private assets cannot be disregarded. The particles upon the surface of the road are pulverised by the incessant grinding of these juggernaut wheels.'

There was a lot more in similar vein, followed by full correspondence columns for weeks on the subject, mostly anti-road vehicle. In fact most of it was directed against the motor car rather than the commercial vehicle, unlike today.

At the opposite end of the spectrum of press opinion, *Punch* was trying to promote the idea of 'the Great Polar Railway' which, unbelievable as it may seem, was to run from the north polar region to the equator — through Britain of course — with junctions off to places of suitable importance in between. Just how this major enterprise was to cross the northern seas, the Alps or the Mediterranean was not explained but the idea had immense popular support at the time.

Nevertheless, that 1878 concession to the agricultural engine encouraged Edwin Foden and his colleagues and competitors, and the industry looked forward to a promised programme of steady reform in road vehicle legislation. But their industry was not strong enough on its own to bring sufficient parliamentary influence to bear. Agriculture had obtained most of what it wanted, and there was no proper road-transport industry as such to speak up for further reform, so oppressive had been previous legislation. Consequently it was not until the petrol-powered motor car came on the scene in some numbers in the 1890s that any major legislative changes were made. Even then it was the farmers who got their claim in first. As the new motor cars began to appear, the farmers argued that the employment of their engines in fields need no longer be restricted to more than 25 yards from the road to avoid frightening road users, because equally alarming machines were on the roads. The 1894 Locomotive and Threshing Engines Act removed that condition, although it still required a man to warn any oncoming horse traffic. But the major step was the Locomotives on Highways Act of 1896, which allowed machines weighing under three tons — called light locomotives — to travel at up to 12 mph without a preceding red flag. Locomotives over that weight could travel at 4 mph only, but, most important of all, they too did not need the red flag.

The road was at last open for serious freight traffic. But seemingly unending years of suppression had left the manufacturing industry in very poor shape to exploit the new-found opportunities, and several years were to pass before freight-carrying wagons became part of the industrial scene in Britain. Those years were, nevertheless, times of great excitement, holding a mixture of success and failure for all involved, not least Edwin Foden and his son, William. Initially any road freight was carried on trailers hauled by road locomotives, which were higher-geared versions of agricultural traction engines, usually fitted with hard rubber tyres or blocks over their iron wheels. But although these machines were impressive and sometimes quite fast, they were too heavy and inefficient to compete seriously with the railway, and new ideas were needed. They were not long in coming.

**Above left** Prospector *in her real working days on the fairground circuit all over Britain in the early 1900s: there was a little less shine on her then than there is now.*

**Left** *Undoubtedly one of the finest preserved Foden engines is Frank Lythgoe's* Prospector, *a splendid showman's engine seen here parading at Ascot in the Historic Jubilee Pageant in May 1977.*

# Chapter 3

# The great industrial engines

In the years immediately following Edwin Foden's entry into partnership with George Hancock, the activity at Elworth Foundry was brisk, though confined to small items. All manner of ironwork from agricultural implements to cast fittings for buildings, machinery, business premises and horse-drawn vehicles was produced. It should be remembered that this was the heyday of cast-iron embellishment that typified much of the Victorian era; houses, lamp-posts, letter-boxes, carts, in fact almost any artefact imaginable, had its share of intricate and often delicate tracery in iron.

Although steam was in the Hancock blood, and in Edwin's too, opportunities to realise fully those instincts were too infrequent for their liking. Their workshop was, for its time, highly mechanised. There were trip hammers, forging hammers, rolls and guillotines, all powered from countershafts driven by a big steam engine. What made it even more impressive was that all this equipment had been made on the spot, as well as the boiler, engine, line-shafting and machines. These were all housed in a substantial building which the bustle of activity made a relatively warm and comfortable workplace, so there was never any trouble in keeping good, skilled men.

All that cast-, wrought- and forged-iron work, with occasional steel fabrication too, kept the little factory busy and profitable. But it was steam power that really fascinated the management. A number of steam portable engines were made for both industrial and agricultural use as early as 1862 but, as we have seen, there was little possibility of expansion in that field because of the brutally restrictive legislation aimed against such engines. The two men turned their attention to 'fixed' engines, therefore; engines for use in industrial premises which had their own restrictions and safety conditions to be sure, but which were by no means as onerous as the ones which bound the agricultural and transport industries. Occasionally they would make a small mill engine for a local sawmill or textile mill in nearby Lancashire. But this was quite a difficult market to penetrate, as most of the mill machinery manufacturers had their own arrangements for steam power supply, either within their own organisations or under contracts with specialist mill engine makers, who set themselves up close to the mills themselves.

It was not until after George Hancock retired from the business that the first big breakthrough came. Despite the trading difficulties, Edwin Foden had made a good reputation for himself and his firm as sound engineers and good practitioners in steam. Almost all the work done up to the early 1870s was for local concerns in the Cheshire, North Staffordshire, South Lancashire, Wirral and Flint areas, a district which had always been, and still remains, distinct from the surrounding regions in its way of life, construction and business methods, and in its attitudes to the outside world. Within this large community Edwin Foden was, at the age of 30, a well-known businessman, with friends and contacts throughout the district. Consequently, when a new wire-drawing mill was built at Warrington in 1872, to supply the growing demand for all kinds of new household and industrial hardware, the

*The great industrial engines*

owners asked Edwin Foden if he could build them a steam power unit that would provide the considerable mechanical forces needed to draw wire from billets of steel. A chance like this was just what Edwin had been looking for. Working from data supplied by the customer, the Whitecross Wire Company, he designed a horizontal engine, with a 20-inch diameter cylinder and a 30-inch throw on the crankshaft. This was before the compound engine principle had been developed, so it was a simple single-expansion engine, with a flywheel about 16 feet in diameter. The engine was carefully built up at Elworth, run from the existing boiler at the foundry, and thoroughly tested with the customers present at the tests. They pronounced themselves satisfied, and the whole machine was dismantled for transportation to Warrington, about 25 miles north. It was now that George Hancock's wisdom in siting his works close to the railway became clear. The heavy engine — its flywheel alone weighed over eight tons — was moved quite easily in pieces on to flat-bed wagons in the Elworth railway sidings, then off up the line to Warrington, where the new wire-drawing mill was also sited close to the railway. Included in the price to the Whitecross Wire Company was the item 'carriage by railway to Warrington, £16 12s 9d'. Edwin was as delighted with the job as were the customers: he was in the steam-power business on a big scale, and he liked the feeling.

A little later that same year, word of the success of the Warrington engine reached the ears of the management of the Hill Top Colliery, at Longton, about 20 miles south-east of Elworth, just over the county border in Staffordshire. It was only a small colliery even by the standards of the 1870s, but this new order was a significant one for the Elworth Foundry, because it proved to all the men who worked there that they were becoming well known for their workmanship and skill. We will see in Chapter 10 that employees and employer worked very much as an integrated team. To be sure Edwin Foden was the boss, there was never any doubt about that, but he regarded his employees as indispensable members of his team, and treated them as gentlemen and artisans, not just as labour. Consequently, there was an air of considerable pride and excitement within the works when the order came in from the Hill Top Colliery, and Edwin Foden and his foreman-fitter Fred Mason went to Staffordshire to study the existing arrangements in collieries for winding cages up and down the pit shafts.

The two men decided that great power was not necessarily the top priority, but instead smooth speed control, and above all gentle and precise stopping and starting of the engine, that were important. It had been common practice for winding engines to be built primarily to lift coal to the surface, with the transport of the men at the end of each shift being of secondary importance. As a result, the average steam winding-gear was somewhat violent in its motion and, when the cages were lowered into the pits, the first 20 or 30 feet was almost a free fall as the winding-rope reeled off a rapidly accelerating drum. It is a reflection of Edwin Foden's wider concern with men and their work, and not merely with the convenience of machines and production which had marked much of the industrial revolution to that date, that his first colliery winding-engine was extremely smooth and controllable. At the same time it gave an unprecedented standard of hoisting in the Hill Top Colliery which remained unsurpassed until electric winding was introduced.

But his genius for engineering was also seen, when it was found that his engine consumed considerably less fuel than others of its kind. In one of the few recorded comments from that era, he said in a letter to his first son William, then at school, 'It is not only in the harnessing of the immense and all-powerful energy of steam that my attentions are directed, but in the controlling of the power so aroused to the utmost degree of precision. Furthermore it is of little consequence producing a great deal of power through steam if it cannot be so done in a most economical and efficient manner so that the cost to the user of the invention does not improve considerably on that to which he is already accustomed'. This, perhaps, was one of the main secrets of his success. Economy of effort meant economy of cost in his style of engineering and, in an industrialised society, economy of cost was a factor that took precedence over most others. Unfortunately that precedence was altogether too marked in some industries.

The Longton colliery engine-set was in fact a

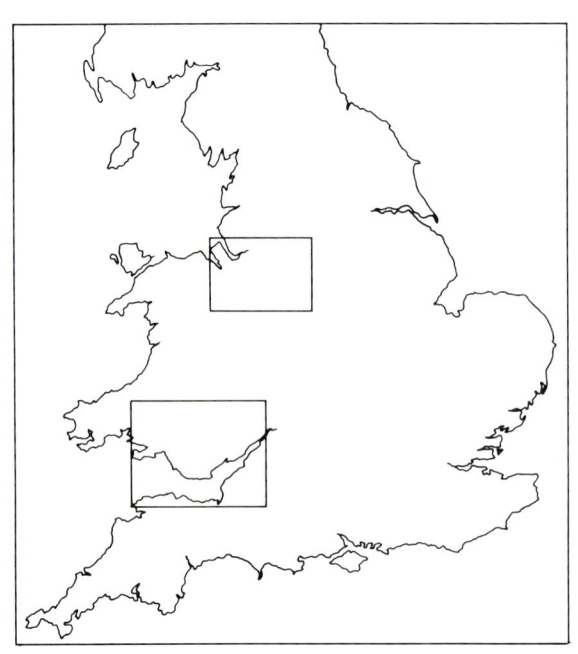

pair of single-cylinder engines in horizontal configuration, smaller than the Warrington engine, with pistons 16 inches in diameter and a stroke of 24 inches. Following the success of that venture, a number of quite small steam-engines was built, including some very compact units built in one assembly with a boiler, and used in small businesses like printing shops and warehouses, where there was little space available. Most of these engines were fired by coal or coke, though a few could by fed with oil fuel. They were, in fact, very similar to the successful threshing-machine engines.

The next big engine order came in September 1875, when the Stour Valley Company, who engaged in the mining of shale and coal, and a

*These maps show the location of the principal scenes of Foden activity during the last 20 years of the 19th century, with (left) their geographical position in England.*

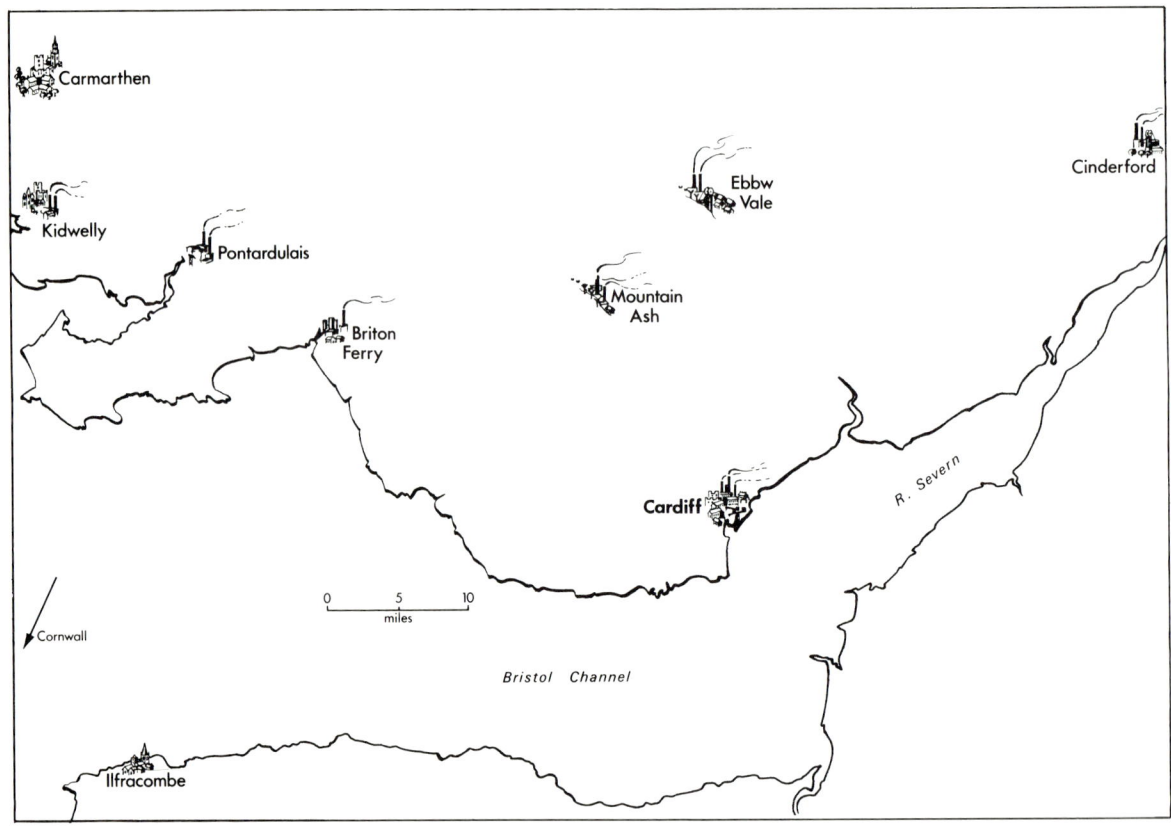

*The great industrial engines*

variety of associated trades like coking and oil extraction, ordered a pair of big winding engines for the new mine they were sinking in Worcestershire, part of the 'Black Country'. Not only was this the largest engine to be built at Elworth up to that time, but the erection site was the farthest away from the works. The cylinders were no less than 22 inches in diameter, and their stroke was 48 inches; a really powerful engine, but no less smooth and economical than the Longton engines. This pair cost the Stour Valley Company £650, including 'carriage and fixing' and also railway fares of 'seventeen shillings and fourpence' for the men who went to erect the engine on the site — Edwin Foden himself and Fred Mason. The Stour Valley Company continued as a regular Foden customer well into the 20th century.

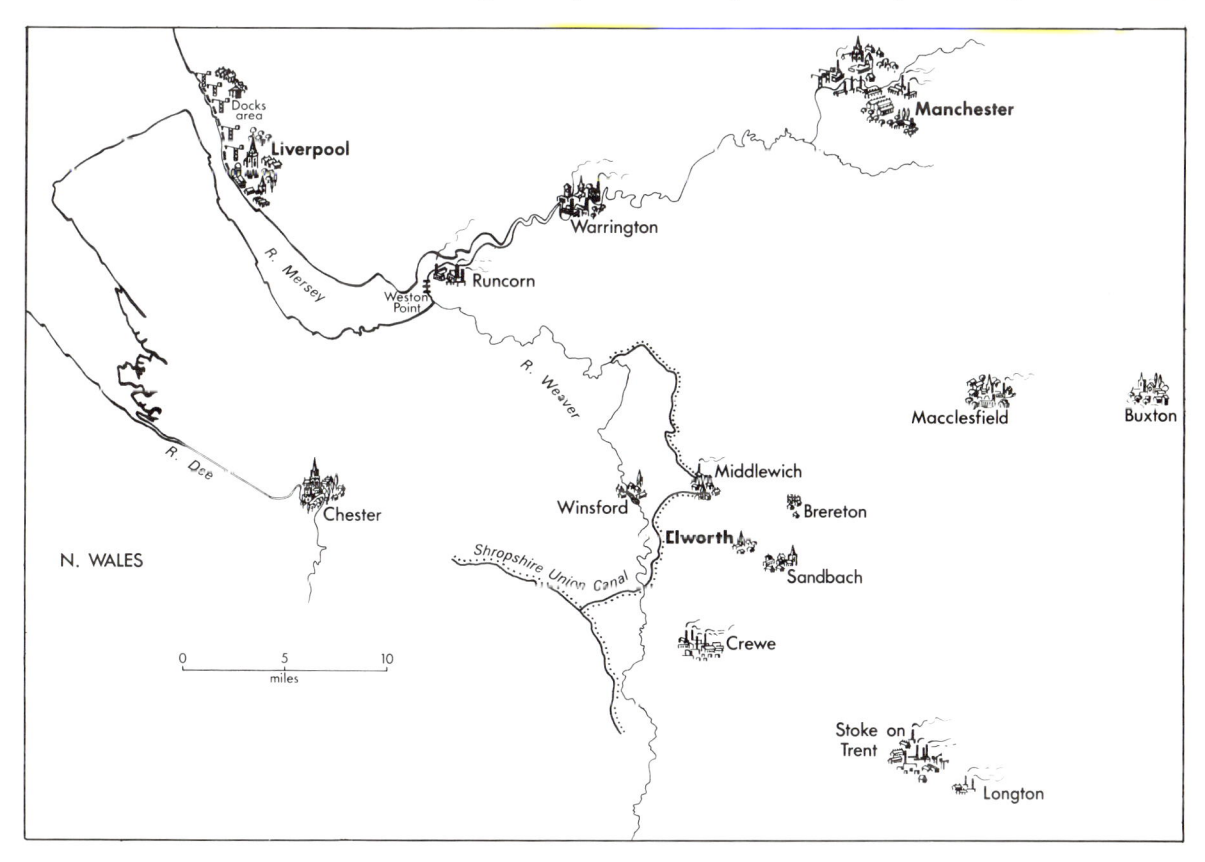

Coal-mining was by then one of the major industries of Britain, and 'King Coal' was the key to much of the success that Victorian engineering enjoyed both at home and overseas. Much of the coal was mined in Lancashire, Yorkshire, Nottinghamshire and Kent, but the most sought-after coal, suitable for big steam-ships, was in some of the deepest pits in South Wales. There, too, pit owners were conscious of the need for economy, and any measure that would enable a more efficient or economical mining operation was greeted with interest. So it was that the efficiency of the Foden winding-engines came to the ears of certain mine owners in South Wales, many miles from the small town and foundry of Elworth. A pair of big winding engines was being built for Hulme colliery, Longton, at the time, but the pit owners there agreed to let their engines go to South Wales after an enquiry from the Ebbw Vale Iron and Coal Company had led to a demand for very rapid delivery. These were built at the end of 1875, and marked the beginning of a long association between Foden and South Wales. The contract price for that pair of engines was £635, which seems amazingly cheap today for massive engines with 18-inch diameter cylinders, and 34-inch stroke, each developing over 700 hp.

That order was followed immediately by another for four engines of similar size, and at an agreed price of £1,020 it was the biggest single item of business that had ever come into the little works in Cheshire. But this record did not stand for long — not that the size of any single invoice was considered a matter of great consequence at the time. In December of 1875 a new mill was built in South Wales by the Williams Jones Jute Sack Company of Cardiff. It was one of the largest jute mills in Europe and they needed a big, powerful engine to drive the line-shafting on several floors, which in turn would drive all the sack-making machinery. They had heard of the big new engines up at Ebbw Vale, and had been very impressed when they saw them. Consequently it was to Elworth that they came to buy their new mill engine: the most powerful and expensive engine that had ever been built at the works. It was a high-pressure engine, with 16-inch bore and 36-inch stroke; the long-stroke configuration was to give the high torque output necessary for driving a large amount of line-shafting. The engine cost £1,757; it developed almost 900 hp and, from all accounts, the owners were delighted with the results when it was installed in the spring of 1876. Since it weighed well over 60 tons all told, its carriage to Cardiff and erection in the new mill was a major task.

As the industrialisation of Britain progressed, the demand for food in convenient packaging developed: the process of preserving food of all kinds in tin cans had spectacular and far-reaching social implications, revolutionising the life style of the man in the street. It brought with it the need for new steel processing plants to meet the increased demands for the tin plate, from which cans were made. (Tin plate is steel rolled to a uniform thickness when hot, then coated with tin or tin alloy to prevent corrosion.) The rolling of the steel was done by passing billets of steel backwards and forwards through several sets of rollers until it was of a specified thickness. This in turn required an engine that could run equally well in both directions, and provide a great deal of power for the rolls.

One of the companies to use the system on a large scale was the Albion Steel Company, Mountain Ash, and for them Edwin Foden designed a pair of large reversing engines with 40-inch cylinder bore and 54-inch stroke. A special valve gear was designed so that the rolling-mill operators could, by moving a single lever, stop the engine precisely at the right moment and set it off in the opposite direction as the billet rolled back and forwards through the mill. Two similar engines were built for the long-established Kidwelly Tinplate Co in West Wales, replacing in 1877 the huge waterwheel that had powered Britain's oldest tin plate works for over 70 years. This was one of the last big single-expansion engine-sets made at Elworth, although some smaller ones were still being produced well after that date, including one for Harrison Saw Mills, in Chester, a small engine with a 10-inch by 24-inch cylinder, costing a mere £130 8s 7d.

As the Kidwelly works was one of the oldest in Britain, and the engines there are still in existence, it is perhaps appropriate to look a little closer at that particular installation, as an illustration of the part that Elworth Foundry played in the improved industrialisation of Britain. The very first recorded production of tin plate in Britain is ascribed to John Hanbury of Pontypool, and soon afterwards Charles Gwynn erected a rolling mill on the banks of the Gwendraeth River just outside Kidwelly. The year was 1737; bars of iron were brought down from the many forges in South Wales that existed at that time. The enormous torque generated by the waterwheel was more than adequate to roll the bars out into sheets which, after 'pickling' or de-greasing, were dipped in molten tin, shipped-in across the Bristol Channel from the mines in nearby Cornwall. The resulting product was called tin plate and, being resistant to corrosion, it was in demand in an increasingly industrial society for pots, kettles, lamp reflectors, and other household artefacts. At that time, of course, the canning of food was an unheard-of use for the material, and in any case the relatively crude quality of the tin plate of that era would have made it unsuitable for the process.

The fortunes of the Kidwelly works ebbed and flowed considerably over the years, with a number of owners and at least one extensive rebuild. But in 1858 Jacob Chivers, a successful industrialist from Gloucestershire, bought the works and set

*The great industrial engines*

*All that remains of the Kidwelly tin plate works is the two Foden engines in the skeleton of the rolling mill, the two boiler stacks nearby, and another engine house containing a horizontal engine.*

about modernising it. For a while he still used water power, but in 1877 he introduced steam power with the two Foden engines that are still there to this day. These two engines drove nine sets of rolls, and the quality of the tin plate was extremely good; good enough, in fact, for use in food-canning, which was by then a new but fast-evolving industry.

Chivers raised the whole standard of living in the area, not only by paying well for a good day's work but also by providing houses for his 300-strong workforce — a very unusual practice in the 19th century. A strong friendship was struck up between Jacob's son, young Thomas Chivers, and Edwin Foden, who worked on-site himself while the engines were being erected. In fact the Chivers family had a long association with the Fodens, and well into the middle of the 20th century Chivers acted as Foden agents in Gloucestershire and Wiltshire. That the two families were convinced of the benefits of large-scale power in industry cannot be denied but, unlike many others, they adopted a paternal attitude towards their employees that stood as an example to industrialists throughout the country. The enormous growth in productivity at Kidwelly after the steam engines were installed is illustrated by the fact that, with water power, 20 boxes of tin plate a day was considered a good output. But the steam powered plant was easily capable of quadrupling that figure, and sometimes 100 boxes a day were turned out. Women worked there as well as men, doing jobs such as cropping the plates square on power-operated guillotines, and 'pickling' the rolled plates before they were coated with tin.

Chivers sold the plant in 1888, although retaining a seat on the board. Unhappily, not long after that, there was a slump in the tin plate business due to international tariff problems, and the works closed for nearly three years. Many of the Kidwelly craftsmen were obliged to emigrate to the USA to apply their skills over there. But by the time Edwin Foden was producing his early steam wagons in 1901, Kidwelly was enjoying prosperous times once more and the giant single-expansion engines were once again turning the rolls and producing high quality tin plate. They continued to do so until 1938, when more modern mills finally tolled the death knell for the 19th century engineering that had served for so long.

The engines are just about all that remains of the plant today, but fortunately they are to be preserved as an industrial monument. Even in the present age, with all its marvels of science and engineering, the awesome mass of the Kidwelly engines cannot fail to impress visitors. How much more impressive they must have seemed some 100 years ago!

Impressive and successful though the 'fixed' single-expansion engines had proved during the

*The south-end engine still has its steam trunking and all the mill-drive gear. One of the guillotines used to crop the rolled steel can be seen in the foreground.*

few short years of their production, Edwin Foden was dissatisfied. He disliked the idea of all that exhaust steam containing, he thought, a large quantity of useful energy, literally going up the chimney. Brunel had developed the principle of the compound engine for use on his fast railway engines to the West Country, as had Stephenson elsewhere in Britain, and although the problems of making it work with a fixed engine were rather different there was no reason why it should not provide a solution. Foden was acutely conscious of a locomotive called the Webb compound, built in Crewe, which had two high-pressure cylinders and a low-pressure one between them which as often as not revolved its driving wheels in opposite directions when attempting to start.

A great deal of experimenting was done at Elworth with valve gears, exhaust arrangements and relative sizes of cylinders, and soon the answers emerged rather more successfully than at Crewe. In February of 1877 Edwin Foden completed his first commercially successful compound engine. In comparison with the giant machines he had been building for South Wales it was a tiny engine indeed, with a high-pressure cylinder diameter of only 9 inches, exhausting into a 16-inch low-pressure cylinder. The engine was sold to Mr Mason of Calverley Mill, in West Yorkshire. While this engine was being evaluated in industrial service, and its fuel consumption carefully monitored, the firm of Jacob Chivers at Cinderford, Gloucestershire, who at the time also owned the Kidwelly tin plate works, enquired about a powerful mill engine. It is recorded that Edwin Foden asked them to wait a little longer so that he could build them a compound engine with higher efficiency, but Chivers was adamant, so the very last big Foden single-expansion engine was built. It was unusual in that its stroke of 42 inches was less than the bore (43 inches), a configuration usually reserved for high-speed marine engines. But this was no high-speed engine with its 40-foot diameter flywheel weighing 32 tons. Foden was not very happy with it, and he vowed never to make any more large single-expansion engines. The future was in the compound type.

It was opportune that, at about that time, the firm of Stubbs Brothers was engaged in expanding the commercial traffic on the extensive canal system that crosses Cheshire into neighbouring

*The great industrial engines*

counties and joins up with the Mersey and Weaver rivers, and other important waterways. Stubbs wanted engines to power a new barge-haulage tug on the canal at Winsford and approached Foden to build them. The first was almost a model, with only 5½ inches high-pressure cylinder diameter, and a pair of engines with a boiler was supplied for £151 19s 5d. Immediately a larger set was ordered and within two months it had been supplied. These were bigger, with 14-inch high-pressure cylinders, supplied complete with couplings and shafts for the propellers. It was the beginning of a long and successful relationship between Edwin Foden and the Stubbs brothers, who seemed to be a Victorian equivalent of the modern 'wheeler dealers', as they had interests in a wide variety of activity. One of their largest purchases was a special brine-pumping engine, used in extracting salt from beneath the Cheshire plains. This huge machine had a 10-foot stroke to its piston, acting directly on the pump crosshead. They also bought one of the first twin-cylinder compound road locomotives to be built at Elworth, followed later by several others.

But if Stubbs Brothers were among the most prolific Foden customers, the strangest must surely have been Clarke & Co of Weston Point, Runcorn. Clarkes were basically marine engineers and dealers, and they were involved in all kinds of typical waterfront activity like lighterage, chandlery, storage, ship repairs and so on. In 1883 they built a twin-screw steam tug for a Liverpool company, and the vessel was named *Clarissa*. Edwin Foden was approached to build the engines for *Clarissa,* and after some thought he consented. The power units were vertical compounds, with high and low-pressure cylinders of 18 inches and 34 inches respectively, and, when delivered in April 1883, they cost £2,263 for the pair. This was one of several sets of engines made for Clarkes, but the strangest job of all was building the hull of a twin-screw steamer, the *Navigator,* for an agreed price of £3,578 10s 2d. The massive hull weighed about 140 tons as a bare shell, and was built in 1884. Just how they moved it from Elworth to Runcorn is not recorded. What probably happened was that it was moved overland to Middlewich, a distance of some five miles, or possibly to Winsford, to where it could be floated on the Shropshire Union canal in water sufficiently deep and wide for it, and thence to the Mersey at Runcorn. The smaller canal that branches off to Elworth was certainly not sufficiently deep or wide for a vessel of that size, and the overland haulage job must have been a spectacular undertaking all those years ago. The biggest job completed for Clarkes at Runcorn was a powerful set of high-speed compound engines made in 1888, with a very short stroke of 19 inches, compared with the 16-and 30-inch bores. The engines alone cost £2,827, and the whole installation with shafts, boilers, piping, controls and everything else netted the biggest sum yet for the Elworth works, £6,756 in all. In the late 19th century this was a great deal of money.

While all that was happening, the activity in South Wales was not neglected. The compound engine had applications in the steel and tin plate

*The engine at the northern end of the plant, now shorn of its steam trunking, looks magnificently simple with its 40-ft built-up flywheel and cylinder supported on four cast-iron stanchions.*

industries as well as in ships, and a series of big compounds was built for the steel industry. The Clayton Tinplate Company at Pontardulais had one of the largest engines, an inverted twin compound. The name is misleading because the engine does not appear to be upside down. Most big engines before that time had the cylinder at ground level with the connecting rod reaching upwards to a beam, on the other end of which was the driving rod to whatever machine was attached. These new engines had the cylinder high in the air with the connecting rod reaching down to a low-level crankshaft, hence the description 'inverted'. The flywheel was made in four sections and weighed 35 tons.

Still not satisfied with the efficiency of the compound engine, Foden made these later compounds as 'condensing engines'. Here the exhaust steam was condensed by passing it through a heat exchanger, so that the heat it contained was transferred to water that was to be fed to the boiler, a process which saved a valuable 2-3 per cent of the overall heat energy input. The Vernon Tinplate Company took delivery of an even more powerful set in 1892, a pair of 28/50-inch compound condensing engines, with all four cylinders laid horizontally and this was installed in a new rolling mill at Briton Ferry in South Wales. In the following year, Mr Foden visited the site, and made a note in his diary that '...this magnificent engine is consuming no more than 35 or 40 tons of cheap slack (broken coal and dust) at five shillings a ton, and in return producing a good 2000 cases of top quality tin plate each week. It is a most effective machine'. There were other factors to be entered into the economic equation of course, not just the straight swop of 40 tons of slack against 2000 cases of tin plate, but the point was made that these later Foden fixed engines were among the best and most effective of their kind.

But, spectacular though the South Wales business may have been, it was not always profitable. The Elworth Foundry became a substantial shareholder in the Ebbw Vale Iron and Coal Company, as Edwin Foden found himself obliged to take shares instead of cash for some of his transactions. The last straw with the Ebbw Vale company came in 1893 when a small winch engine was supplied. The original contract price was £395, but so many changes were demanded before it was finally installed that the cost finished at £596 8s 4½d. The customer insisted on the contract price being adhered to, and it seems that at that point Edwin lost patience with his South Wales friends, as no further major work was done in that area.

Furthermore, it was almost the end of the 'fixed engine' era. The wire mill at Warrington that had been one of the first customers, and which now belonged to the Sankey Company, came back for a second engine for a new wire mill, which cost them a bargain £616, and a very small engine was made for a Mr Wallace at Stud Green, Sandbach for a mere £52 11s 2d. One or two more mine engines went to the Longton area of Staffordshire, and an engine was supplied to Lea Brothers, corn millers at Brereton, just a few miles from the works. It was later rebuilt and modified as a compound on-site, any parts that needed major attention being taken by road to Elworth, and it remained on its original site until about 1971 when it quietly vanished, presumably broken up for scrap.

But by the late 1890s, Edwin Foden's mind was on other things. Legislation had changed attitudes towards road-going steamers, although public opinion seemed to be scarcely less hostile. A great deal of technology and know-how had been acquired in the 20 years or so during which the great industrial engines had been built, and much of this was already being applied in Edwin's fertile mind to the problems of making a motive steam machine with a greater efficiency than anyone else's. That meant a considerable change of emphasis at Elworth, but the great engines made in that small works had not only earned fame — and considerable fortune, too — for those who made them, but had made a substantial contribution to industrial progress in Britain. However, greater progress was still to come, though nobody then knew just how great.

*Fodens' involvement with both agriculture and industry was combined in the picturesque corn mill at Brereton in Cheshire. The mill engine (inset) was on the first floor at the right of the building. One of the smaller industrial engines, it was still in position in the early 1970s but has now disappeared.*

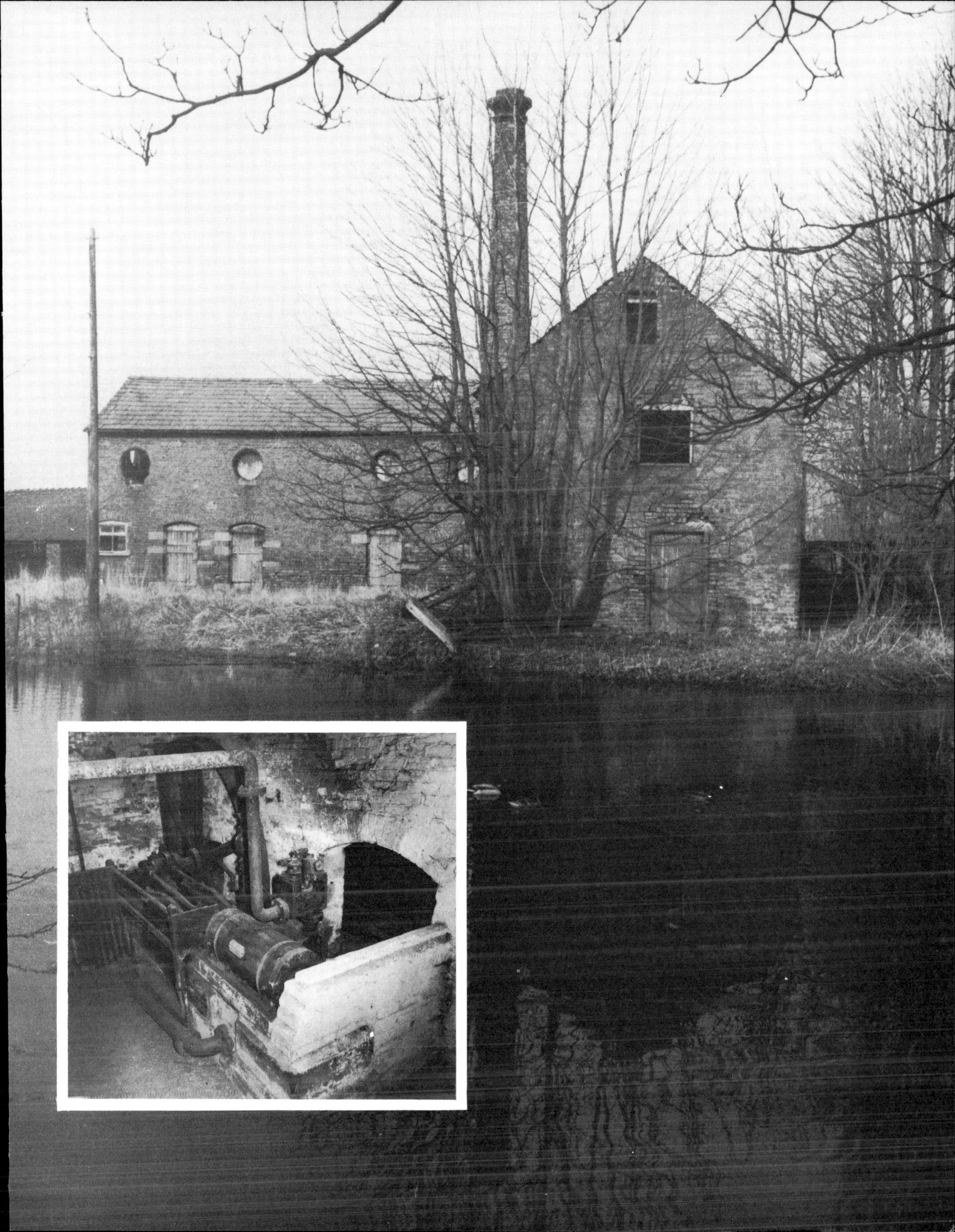

# Chapter 4

# Working with the Victorians

Much of the 19th-century development of industrial Britain, was accompanied by appalling standards of welfare, tyrannical management, poverty-line wages, intolerably long hours in the factories, and child exploitation. Not all industries were as bad, but few were as oppressive as the cotton industry, for example. The diary of one William Varley, a Lancashire weaver, contains an entry in 1826 ' ... there is a great many people that is poorly about this time and well there may be, what with hard work and mean food. But there are many more without work and what must become of them? They must lie down and starve to death for all I know. If they should beg I know of none that would give anything, and if they should rob or plunder they have the soldiers ready to give them their "last supper".' In February 1827 Varley wrote, 'The weather is uncommon rough ... sickness and disorders of different kinds prevail very much. The pox and the measles takes off the children by two or three from a house, and well may they die for there is no aid for them, the times is no better for the poor. Hunger and cold are our true companions.' Later that year he wrote, 'This day there is a general lowering of wages ... the poor weaver may now go to despair indeed.'

Faced with this abject misery in the mills, the newspapers of the day did little to help, hardly even acknowledging the problem. For example, the *Manchester Mercury* in an editorial in May 1841 — the year Edwin Foden was born — said, 'There is, there can be, no other test of the intrinsic utility of a new machine than whether it effects better or more cheaply the purpose of that which has previously been in use.' The opposite end of the spectrum of opinion is shown in a letter by one Francis Place, a labour leader who tried for many years to improve working conditions in the northwest, particularly in the cotton mills. His view was that '... the manufacturer looks only to his immediate profit and cares little or nothing for what may be the state of trade hereafter, or what may be the fortunes of his successors. The struggle is one of strength alone, and the weaker may go to the wall. Whatever the workpeople gain or retain is gained or retained by power alone, and will always be thus'.

Little wonder, then, in the middle of the 19th century organised bands of dissidents raided factories and mills in Lancashire, Nottinghamshire and even Staffordshire, breaking machines which stood for the exploitation of the workers by management. The bands were not, as is popularly supposed, hunger-crazed rabbles smashing indiscriminately, but carefully organised revolutionaries — counter-revolutionaries perhaps — with specific targets. How else can one explain the smashing of every weaving-loom in several mills in Lancashire, while not one single spinning machine was touched? But despite this inevitable reaction to starvation wages and over-long working hours, and the riots and bloodshed which continued well into the second half of the century, conditions in many industries remained deplorable. Undoubtedly the cotton industry was one of the worst, and coalmining little better. The railways, on the other hand, effected a general raising of standards for its employees, many of whom lived in railway-owned houses which were small but strong and dry, and numerous examples of these can still be seen in railway towns like Swindon and Crewe. Somewhere in the middle of the 'league table' was the engineering industry, which was very much more fragmented than that of cotton, coal or railways, with a large number of small concerns like the Elworth Foundry, and relatively few large companies.

It was against this social background that young Edwin Foden went to join George Hancock in the little factory at Elworth. His father, a staunch Methodist, was a shoe manufacturer and general dealer in the village of Smallwood, three miles from Elworth, and travelled the district regularly, sometimes taking his son with him. Another branch of the family were clockmakers. Edwin was consequently aware of contrasting working conditions in the industry. He had seen the neat rows of railway houses down the road in Crewe, and he had observed the pitiful squalor of less fortunate men and their families in mining villages to the east and north, and the despair of those in the textile towns.

It so happened that conditions at the Hancock works were reasonable when Edwin Foden became a partner in 1866. It was at least warm, with the furnaces for the molten iron, and the engine boiler that provided the power for the trip hammers, forging hammers and other machines. The roof, where there was a roof instead of a hole to let out the fumes and smoke, was sound. The floors were earth — there is little else a foundry floor can be made of — and the men worked on average 11 hours a day, six days a week, for about 21 shillings. That was for the first-class 'tradesmen' or artisans. Those with lesser skills would get about 18 shillings, and apprentice boys were paid from three shillings and sixpence to five shillings and sixpence, depending on their seniority and how long they had worked there. These rates of pay did at least provide food and rent for the men, and consequently they considered themselves better off than many other workers, as indeed they were.

Foden had already recorded his opinion on how machinery and steam power should be developed, and it was not dissimilar from the view expressed in the *Manchester Mercury*. But the big difference was that Foden, perhaps because of his strict Methodist upbringing, did not propose to extract that ultimate efficiency from his machines at the expense of the people who worked them. His was a more humane approach. To Edwin Foden, the men who worked with him were not merely units of labour, they were people. Make no mistake, there was never any doubt that he was the boss, but he gradually bred a unique spirit among his small team of between 30 and 35 men and boys. His philosophy was that, if a man has the skills to do a job, that job would be done better and perhaps quicker if the man enjoys the confidence and friendship of his superior. He also believed in confiding in his workforce his future plans and intentions, so that they felt part of the overall operation and could contribute accordingly, no matter in how small a degree. In the last quarter of the 20th century, this doctrine may seem trite and unsurprising, but 100 years ago it was not only new and advanced thinking — what we would now call a sophisticated labour relations attitude — but it also worked admirably.

Just how closely Edwin Foden worked with his men can be seen from the number of occasions on which both he and his head fitter, Fred Mason, went off together to survey a job before setting out to build an engine, and then returned a few months later to erect the engine on-site. Few bosses at that time would be seen in overalls and bowler hat perched high on a scaffold or building frame assembling their products with all the care and skill of the leading artisans. But that is where you would find Edwin Foden, for not only was he a good manager, he also had all the skills of the engineering trade, and there was little that his men could do that he could not do with equal ability. Throughout his career he was inclined to peer critically over the shoulder of an employee at whatever work was in progress, then politely take the file or scraper from the man and proceed to do the job. 'Do it this way lad, then tha'll do it reet', he would say. Nobody complained, perhaps because Edwin could probably do the job better anyway. This brought him immense respect and loyalty — one of the major strengths of his little company. Invariably, on the erection of a big fixed engine, like those at Hill Top Colliery and the Cardiff Jute Sack Mill, he would take along at least one apprentice to act as fetcher and carrier of the numerous bits and pieces required to assemble an engine, and also to learn the wider implications of the type of engineering that they practised at

Elworth Foundry.

His concern with his men was not confined to working hours either. There are records of several occasions when he visited the homes of men who worked at Elworth where a wife or a child was sick, with words of comfort and frequently a small gift or a recommendation of a good physician who would help; as often as not the employee would find that any bill would be paid by Elworth Foundry. This was not simply paternalism or philanthropy on Edwin's part, but a positive attitude to the progress of his team. He acknowledged that, if a man's family had sickness, he would not be in a fit condition to dedicate himself 100 per cent to the job in hand. It was consequently as much in his own interest that Foden looked after the well-being of his workers' families, as in the interest of his employees. But his form of Victorian industrial paternalism was none the less a welcome trait in that part of Cheshire.

It was not until near the end of the century that a formal welfare department was established at Elworth. By then the workforce had swelled to over 100 men and it was quite beyond the capabilities of either Edwin or young William to take care of things personally and run the expanding business too. Many lesser men would simply have let things lapse, but the Fodens were not lesser men and they set up a welfare system. By today's standards it was primitive, but in the closing years of the last century, it was looked on with great admiration by those fortunate enough to benefit from it, and by those less fortunate outside the organisation who worked for less enlightened management.

It was undoubtedly the existence of this welfare department that enabled the large expansion of the company, after the successes at the War Office Trials in 1901, to go ahead without undue difficulty in finding adequate skilled labour. There were considerable difficulties in other areas like financial structuring and formulating suitable investment policies, and other responsibilities on a higher management level where there was no previous experience within the company. But there were few problems in attracting the necessary workforce to come to the Elworth Foundry (as many still called it, even though the name was changed to Fodens Ltd

after the restructuring in 1902). Proper wash rooms had been installed well before the turn of the century where those men who did the dirtier jobs could clean up and change, if they so wished, before going home. There were proper facilities for brewing tea within the works, and a first-aid medical centre was established with trained medical help on call by telephone in the event of a workman being injured on the premises.

The first facilities for recreation activities outside working hours appeared at that time, too, beginning with a bowling green and a social club.

*Plans of the proposed North Staffordshire railway siding extensions for Elworth Foundry in 1887 show the extent of the works at that date. The rail extension was intended to make the shipping of heavy engines easier. The works extension shown on the plan was completed the following year.*

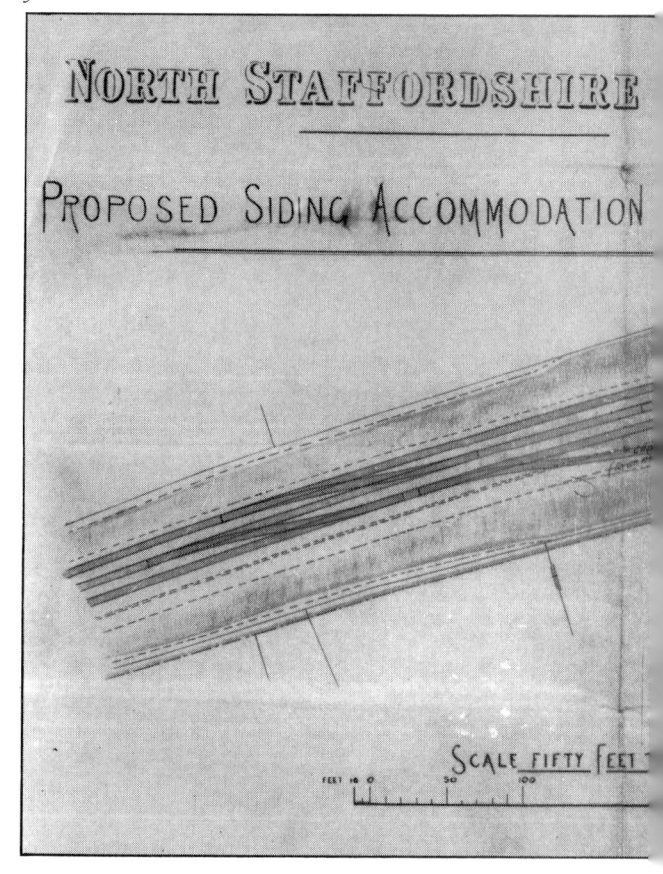

Later on these small facilities grew into much larger concerns: by the early '20s, for example, there were two bowling greens, a tennis club with good courts, billiard rooms, and an excellent club building. The company built considerable numbers of houses for their more senior workers, many of which still stand. The famous Foden Motor Works Band grew from somewhat inauspicious beginnings as an offshoot of this socially orientated activity, with several members of the Foden family among its ranks. The story of the band is told in Chapter 11, but this illustrates how advanced the social awareness was at Elworth when it was first formed in 1902.

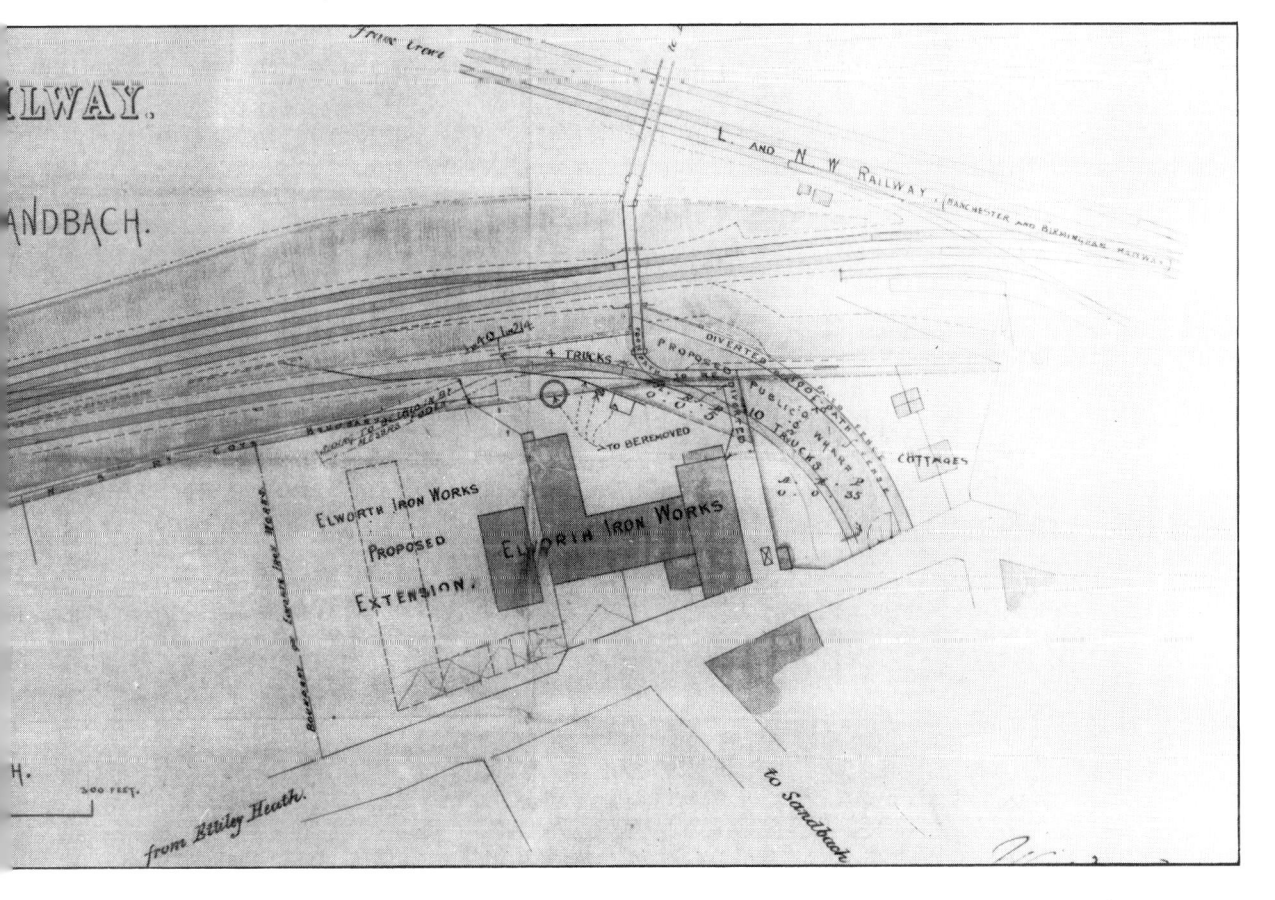

It is difficult to discover when company housing first began. Certainly a small number were built in the 1890s for leading skilled men in the company, and some items in early balance sheets show that there were at least a dozen rented houses as early as 1885, while more were purchased in ones and twos during the same period. It is interesting to note that the provision of company-owned housing for workers is seen as a major social revolution in countries such as Germany in the second half of the 20th century. True, the scale of these modern housing schemes is very large, but the principles were laid in Britain more than a hundred years before, first with the railways and later with enlightened companies like the Elworth Foundry. Indeed, it seems that Edwin Foden moved in good company when he adopted such policies. Certainly a number of the major customers to whom he supplied his giant steam engines looked after their employees in a remarkable fashion, in exchange for total loyalty and, it must be said, an extremely hard working life. When he and Fred Mason were working in South Wales in the 1870s, erecting those big engines, they would find neat rows of modest cottages provided for the main force of workers at places like the tin plate works in Kidwelly, Briton Ferry and Pontardulais, and rather more primitive accommodation in the mining valleys around Ebbw Vale. Foden was nothing if not observant. He would note that where there were excellent houses such as those at Kidwelly, there was great stability among the workforce, and such stability could lead to better efficiency. The cost of the housing was, consequently, a sound investment for the company, quite apart from the ownership of property which was always considered a good investment, regardless of the circumstances.

These ideas on stability and loyalty bore fruit in no uncertain terms. Many of the employees at the Elworth Foundry spent their entire working lives with the company, and in many cases son followed father, and grandson followed son. Such a case is the family of Fred Mason, the foreman fitter of the big fixed-engine days. He worked on right through the golden years of the steam wagons, and was joined by his son, Ernest, who worked for Fodens for an astonishing total of 60 years. Ernest started as an apprentice, became a pattern-maker and eventually chief

*Fred Mason was Edwin Foden's right-hand man, and as foreman fitter he travelled with the boss wherever new engines had to be installed.*

draughtsman. Men of that calibre were invaluable to companies like Foden. The translation of ideas and sketches into properly designed working machines was the responsibility carried by the likes of Ernest Mason. The process would be considered very unscientific today, but it worked. Edwin would discuss his ideas, and what he wanted a particular engine or machine to do, and sometimes notes and sketches might be made on a piece of paper. More often than not it would merely be chalked lines on the shop floor or a piece of boiler plate. Mason would go to work on those ideas, considering every possible way of achieving the aim. Would it be best as a casting or a steel fabrication? Need it be machined? How much would it weigh, and could that be reduced by doing it another way? There was a lot of experience involved, and indeed a lot of trial and error too. But the final products were masterpieces of Victorian engineering, and the standards set by those dramatic times were carried on well into the middle of the 20th century, when more closely defined methods of attacking an engineering problem took over.

The physical working conditions were often less than luxurious at Elworth. A foundry is essentially a rather dirty place, although of course it was usually warm, which was a major item in its favour. However, the workshop was quite well lit, and there was plenty of power available for removing much of the drudgery from metal-working. Nevertheless, the work of building boilers and engines was frequently uncomfortable and painful, with men squeezing inside the boilers or engine structures, a candle in their caps rather like miners, to work on fitting and finishing the interiors. Boilers were particularly difficult to work on, as the space was very limited, and there was a lot of interior finishing to be done. Nevertheless, there was a quality about the Elworth products that enabled them to function for extremely long periods. Some of the fixed engines worked continuously for half a century, a few of them even longer. The Kidwelly engines, for example, erected in 1877, drove the rolls at the tin plate works until just before World War 2, and even then it was the obsolescence of the plant that forced its closure, not the demise of the engines. Quality of engineering went hand in hand with quality of life at Elworth.

## Chapter 5

# Wagons and the War Office Trials

After the so-called 'Emancipation Act' removed the need for the red-flag man in front of every road vehicle, and raised speed limits to 12 mph for vehicles not exceeding three tons unladen weight, there was a spate of commercial vehicle activity all over Britain. Many companies like Taskers, Burrells, and Wallis & Steevens produced road-haulage versions of their agricultural engines, intended to draw one or several trailers, and to a limited degree they were successful. The biggest limitation on the use of these engines was poor control at higher speeds, especially downhill. Brake technology was very primitive at that time, and most steam tractors braked by reversing the flow of steam through their engines, so that the pistons resisted the motion of the vehicle instead of assisting it. Unfortunately, this frequently led to the driving wheels locking and losing directional stability while the other wheels would continue to rotate, resulting in a 'jacknife'.

Readers might think that the jacknife is a modern phenomenon associated with articulated trucks, but it was quite common as long as 80 years ago. Perhaps one of the most publicised jacknife accidents occurred on a commercial vehicle trial in 1906, when a Tasker *Little Giant* drawing a trailer got into difficulties near Bath on the 1 in 10 Bathford Hill. The driver reversed his engine to slow the vehicle, which promptly jacknifed and overturned into a ditch. That the machine was back on the road within 75 minutes, and on its way again with a new front axle and canopy within two hours, speaks volumes for the efforts of its valiant crew, and for the repairability of the Tasker machine. But the basic weakness was always there with this type of haulage method. It is a tribute to the foresight of Edwin and William Foden that, as early as 1892, they had decided that the real answer to road haulage was not the locomotive or tractor with trailers, but a properly designed wagon or lorry — although of course they built many road locomotives before perfecting a wagon that would work properly.

The trouble was that there were no precedents at that time to indicate the best way of going about the task. Should the boiler be vertical or horizontal? If the latter, should it run fore and aft, or across the frame? Which wheels should do the driving and which the steering? All these very basic questions remained unanswered as Edwin and William struggled with their first wagon design. Only three factors were fixed; all the others were variables in the equation that had to be solved. The dimensions of the wheels were laid down by law according to their load-carrying capacity: they could be calculated approximately from the second factor, the three-ton load that the Fodens had fixed for their new machine. The third known factor was that their highly successful compound engine had to be incorporated in the design.

Not surprisingly, finding the ideal solution to the problem required a great deal of trial and error. In fact it was only at the fourth attempt that anything like a workable configuration emerged from the workshop. The first wagon was a very strange device indeed, but it should be remembered that there were no guidelines to follow; the Fodens were pioneers in the field. The engine was mounted vertically at the front of the

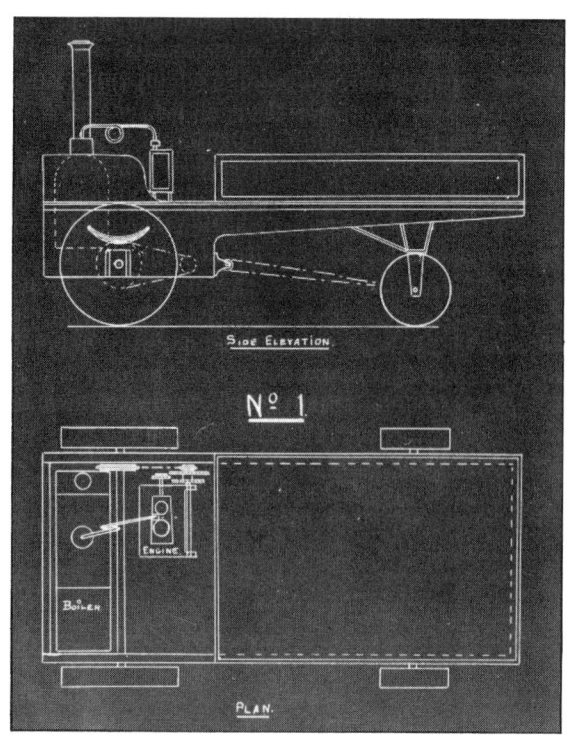

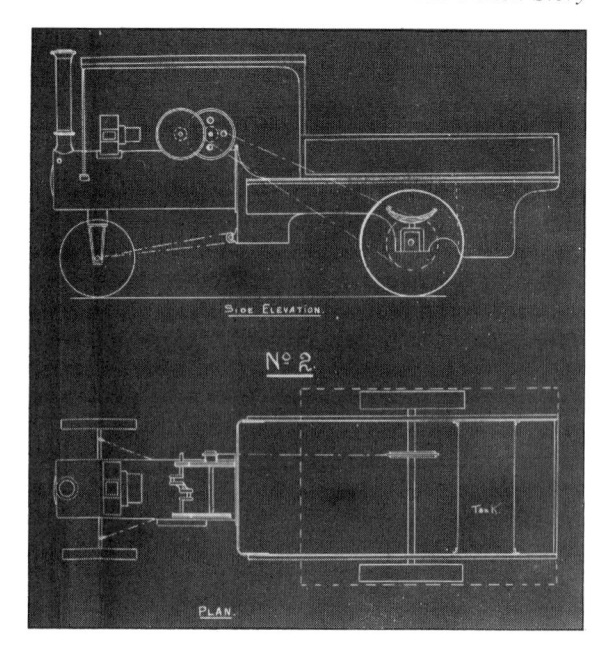

*This unique series of blueprint sketches shows the progression of Edwin Foden's four prototype steam wagons. The main distinguishing feature of the final design was the method of mounting the boiler directly between the frame rails. A photograph of No 4 prototype appears on page 40.*

chassis to drive the front wheels, and the steering gear operated on the back axle. Solid rubber tyres were fitted over the fabricated steel wheels. The wagon made its trial runs at the time of Queen Victoria's Diamond Jubilee, in June 1898. It worked after a fashion, but it was difficult to drive and even more difficult to load so that the wheels carried a proper proportion of the total weight.

However, useful experience had been gained, and a second prototype was built. This experimented with a friction clutch drive, instead of dogs and gears, but was still not the answer, even though it had a loco boiler and rear-axle drive. Time was passing by rapidly; it was 1899 and still there was no solution which wholly satisfied Edwin and William. Prototype number three was nearer the mark. The transmission problem was overcome by using toothed wheels and a long chain to drive the rear axle, while the front axle was used for steering. The engine was still up front but mounted over the top of the boiler, with reduction gearing between it and the chains. This appeared late in 1899. Before the fourth prototype had been designed in the spring of 1900, the War Office announced that they proposed to hold a series of trials for steam vehicles the following year.

This news really put the Fodens into top gear. Day and night they worked, drawing out new layouts and details, discarding some, adopting others, all the time working out the theoretical performance of each part, instead of simply relying on the final practical trial to give them this information. In fact they did not build a complete new wagon, but stripped and extensively modified the existing prototype. The main distinguishing feature, apart from the locomotive-type horizontal boiler with the engine on top, was that the chassis frame rails were bolted to the sides of the boiler and then spread out to pass outside the rear-wheel bearing boxes like a railway wagon. This time the wagon worked well. It was extremely controllable, its turning circle was compact, it carried its rated payload with ease, and coped with severe overload quite happily. The engine by this time had been developed so that both cylinders

Wagons and the War Office Trials

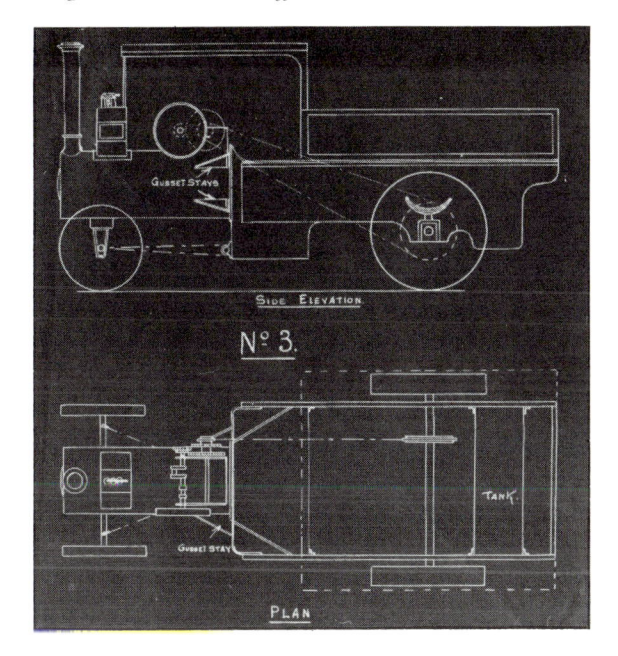

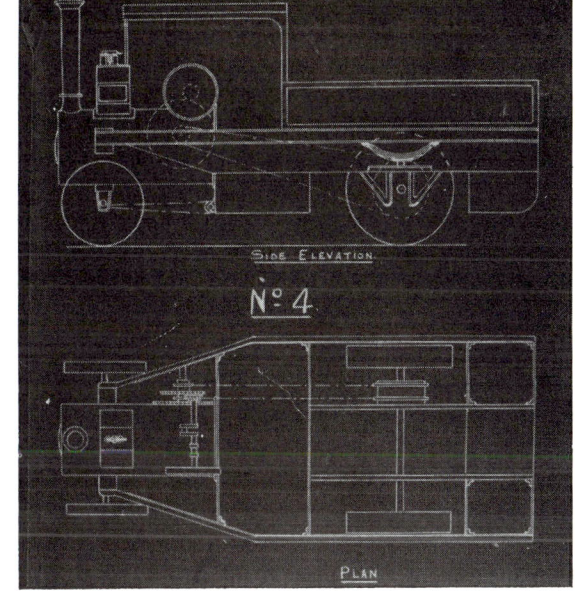

*Diagrams labelled "No. 3" and "No. 4", each showing Side Elevation and Plan views.*

could run on high-pressure steam in an emergency, as well as the usual compound arrangement where the low-pressure cylinder ran on the exhaust steam from its high-pressure neighbour. The valve arrangement ensured that both cylinders developed equal power and so ran smoothly, and this made the Foden engine unique. In the event of a breakdown it could also run on one cylinder. This system gave the wagon a remarkable ability for hill starts and severe pulling jobs. What is more, the fuel consumption throughout the initial test runs was very encouraging indeed. Edwin was satisfied that he had found the answer to his problem. The date was 1901 and it was late summer.

Throughout the period when Edwin and William were developing their new steam wagon, a series of competitions known as the Liverpool Trials was being held in and around that city, organised by the Liverpool Self-propelled Traffic Association. This rather quaint name embraced a group of Liverpool industrialists and merchants who were sufficiently far-sighted to believe that road freight traffic would be of immense benefit to industry generally. They therefore tried to promote it in their area in order that they and their great port should be the first to benefit. Other trials took place in various parts of the country

and the first of several Liverpool Trials took place in 1898, which was the earliest that could be arranged following the 'Emancipation Act'. Young William was keen to have the prototype Foden there, but his father was quite adamant that he was not going to have his machines out in public trials before he was satisfied with them. And he was proved quite right in that decision.

At the first Liverpool Trials all the entries were wagons with vertical boilers and engines under the frame, known to steam men as 'undertypes'. In this respect the layout resembled that of Hancock's steam carriages of 60 years earlier. But two men were not happy with that idea, one of whom was Edwin Foden. The other was one P. J. Parmiter, and he designed a wagon with a horizontal locomotive-type boiler and the single-cylinder engine on top of the boiler. His design was eventually adopted by Mann's Patent Steam Cart and Wagon Company of Leeds, just too late for the first Liverpool Trial, but in time for the second. This type of machine, which became known as the 'overtype', suited road use far better than the existing undertypes for a variety of mainly technical reasons. By dint of exhaustive trial and error, the Fodens had arrived at the same conclusion as Parmiter and Mann at about the same time. The 'Mann Steam Cart' aroused great interest at the Liverpool Trials, and at the third in the series in 1901 it emerged a clear winner. There is no record of whether or not Edwin and William went to Liverpool to see the opposition but, as their premises were less than 30 miles away, it is difficult to see how they could have stayed away. What is recorded is that Edwin had his sights set on the much more important War Office Trials, which would attract nation-wide attention, and that he had no intention of showing his hand until that time. On the other hand, he knew what kind of standards the opposition was likely to achieve from their results in the Liverpool Trials, that opposition comprising Thornycroft, Leyland, Mann, Coulthard (later combined with Leyland), Clarkson and Bayley.

The newly formed War Office Committee on Mechanical Transport had decided that the rapid development of mechanised road transport both in the light passenger-carrying field and in the heavier freight and traction field would represent a powerful strategic force in any hostilities that might arise; and it should be remembered that throughout that period there were few years when a war on some scale or other was not going on. In fact the Boer war was taking place at that very same time. The manufacturers on their part were equally aware that approval of their machine by the War Office was the best possible publicity that they could obtain, and was bound to bring a flow of business from industry as well as from the War Office itself.

The wording of the announcement of the trial was significant. It was published as 'Notice of Competition of Self-Propelled Lorry for Military Purposes'. It was the use of the term 'Lorry' that made these trials different. 'Lorry' was understood to mean a vehicle carrying freight on its own back, not on a trailer. This latter involved a tractor or traction engine, and the insistence on the vehicle being a lorry and not a tractor excluded most of the big steam practitioners like Fowler, Tasker, Ransomes, Burrell, Garrett, Marshall and others. However, the drawing of a trailer by the lorry was demanded under the rules. Compared with the highly detailed and complex conditions laid down for War Office vehicle trials these days, the 1901 rules were refreshingly simple. The prospectus simply read:

**'Rules and Conditions**

1. The conditions laid down in the Notice of Competition, $\underline{Gen.\ No.\ 5}\over 862$ will be strictly adhered to in every particular.

2. Each competitor shall himself make all arrangements for the necessary staff and appliances to work his vehicle or vehicles.

3. All fuel and water will be supplied free of charge by the War Department, and no other will be permitted to be used.

4. Accommodation for the vehicles will be provided by the War Department, and all vehicles shall be stored by night at these depots.

*Wagons and the War Office Trials*

5. Vehicles intended for trial must arrive at the War Department depot not later than 2 pm on Tuesday, 3rd December, 1901.
6. All vehicles, from the time of the commencement of the trials until the completion thereof, shall be considered as in the custody of the Secretary of State for War, and any vehicle removed for any period from that custody without the written authority of the Secretary, Mechanical Transport Committee, shall be ineligible for a prize.
7. Official observers will accompany each vehicle during the trial runs to take notes of behaviour, fuel and water consumption, etc., and no repairs will be permitted without their knowledge and consent.
8. Each competitor shall arrange to have his vehicle or vehicles ready for inspection by the Mechanical Transport Committee, or by any inspector appointed by them, at 10 am on Wednesday, 4th December, 1901, at the War Department depot, when the trials will officially commence.
9. At any time during the trials any vehicle or motor, or part thereof, may be opened up for inspection by the Mechanical Transport Committee or by any inspector appointed by them.
10. Each vehicle will be allotted an official number, which shall be displayed during the continuance of the trials. Boards bearing these numbers will be provided by the War Department for attachment to each vehicle.
11. Ten half-plate unmounted photographs of each vehicle shall be furnished by the competitors not later than 4th December, 1901.
12. A "detention allowance" of 5s per diem will be paid by the Secretary of State for War to one attendant for each lorry for every day during the period that trials of that particular lorry are being carried on.'

Following the successful development of the fourth prototype steam wagon, Edwin Foden set about converting the design to a commercially realistic version. In the main this consisted of making it larger to give acceptable space both in the cab and on the load-carrying area, and it was this version that was entered for the trials. The official starting date of December 4 1901 was deliberately chosen by the committee to make sure that the tracks, lanes and roads over which the vehicles were to run were wet and slippery. There were obviously going to be no short cuts to success in this competition. In fact December 4, which was a Wednesday, was spent in assembling the vehicles and their crews, briefing them on what to do, explaining the rules and penalties, and preparing the wagons after their journeys from various parts of the country. As the trials

**Left** *A modified version of Edwin Foden's third prototype steam wagon formed the basis for the fourth and final prototype, shown here. These trial machines were only about two thirds of the size of the eventual 3-tonner steam wagon.*

**Right** *The War Office Trials wagon embodied numerous improvements on the prototype wagons. This picture shows the machine on completion at Elworth. The unladen weight was comfortably under three tons without water or coal.*

were held in the area around Guildford, Aldershot, Alton, and Odiham in Hampshire — to this day a trials area for military vehicles — many competitors had steamed a long, long way even before the trials began, not least Edwin and William Foden, who had driven their wagon the 160 miles from Elworth without any problems at all. Consequently they were both quietly confident that their beloved machine would do well, regardless of whether or not it beat the other

## List of Competitors

| Offical Number | Name of firm | Engine | Transmission | Fuel |
| --- | --- | --- | --- | --- |
| 1 | Messrs Brown & May Ltd, North Wilts Foundry, Devizes | steam compound 600 rpm | 3-speed, gear | coke |
| 2 | The Creek Street Engineering Co, Deptford, SE | steam 3-cyl radial poppet valve | 1-speed, gear and chains | oil |
| 3 | Messrs Edwin Foden & Co Ltd, Elworth Works, Sandbach | steam compound overtype | 2-speed, gear and chain | coal or coke |
| 4 | Messrs George F. Milnes & Company Ltd, "Motoria", Balderton St, Oxford Street, London | internal-combustion petrol 4-cyl 25 hp | 4-speed, gear | petroleum oil |
| 5 | The Straker Steam Vehicle Company Ltd, 9 Bush Lane, London EC and Bristol | steam compound undertype | 2-speed, gear and chain | coke |
| 6 | *The Thorneycroft Steam Waggon Company Ltd, Steam Waggon Works, Homefield, Chiswick | steam compound undertype | 3-speed, gear with differential lock | coke or oil |
| 7 | *The Thorneycroft Steam Waggon Company Ltd, Steam Waggon Works, Homefield, Chiswick | steam compound rear-mounted | 3-speed with differential lock and articulating axles | coke or oil |
| 8 | The Wantage Engineering Company, Wantage, Berks | steam double-pressure undertype | 2-speed, gear and wire rope | coke (oil or wood) |
| 9 | Messrs Bayleys Ltd, 42 Newington Causeway, SE | | | |
| 10 | Mr Chas. Innes-Baillie, "Ballygunge", Grove Hill Road, SE | Entered, but did not furnish any particulars of vehicle for entry in the programme. | | |
| 11 | Mr J. E. Liardet, 16 Hyde Park Gate, SW | | | |

*This is the actual spelling of the name in the War Office Trials programme although other contemporary and later documents used the spelling Thornycroft. In his personal record of the trials, Edwin Foden spelled it Thornicroft.

wagons, many of which were unknown quantities regarding performance and fuel consumption. There were 11 vehicles entered in the trials, some of them private entries by enthusiastic amateurs, but also including Thornycroft, Milnes and Straker with fully developed professionally built machines, one of which — the Milnes — was petrol- rather than steam-powered. The Foden, however, had by far the most advanced engine design, and the chassis design had been developed considerably on the trials machine compared with the original prototype. They all carried three tons payload on the wagons themselves, and for the majority of the road trials they pulled a further two tons on a trailer. So the all-up weight on poor road surfaces was considerable, in fact about 12-13 tons in all in the case of the Foden and Thornycroft machines.

The first day's test, on Thursday December 5, was a straightforward run, much of it on rough roads, with a number of tricky hills to negotiate both up and down; the total distance amounted to exactly 30 miles. Despite drizzly weather which made the surfaces tricky, the Foden completed the course in 5 hours 45 minutes on 419 lbs of coal and 227 gallons of water. That was seven minutes faster than its nearest rival, the Thornycroft, but the Thornycroft used 146 lbs of coal more than the Foden machine.

Friday's run was aimed at seeing how well the wagons could cover long distances at relatively high speeds, and the hills were less severe than on the previous day. The Foden was fastest over the 30 miles with 4 hours 45 minutes — almost 6½ mph — including all stops, which was unheard of with that weight at that time. Nearest was again the Thornycroft, fully 36 minutes behind, and such were the efforts of the Thornycroft crew to keep up with the Foden that they stoked 638 lbs of coal against the Foden's 359 lbs. The Straker was slower still but used only 500 lbs of fuel. Edwin's ingenious engine design was showing its mettle, providing high power when it was needed for hill-climbing, but throttling back to very economical cruising on more level ground, and in effect running two cylinders on one cylinder's ration of steam. Of course the skill of the crew played a big part, and none were more skilled than the Fodens, father and son.

The weekend saw no further competition, but the national press, which had been watching the trials with interest, reported its findings. Saturday's *Daily Chronicle* commented, 'The Foden Company's steam car which was the first to return on Thursday's trial was first again yesterday, improving on the previous day's time by about an hour.' Edwin and William read the press reports with a great deal of satisfaction, and spent the weekend resting. The second week

*Layout drawing of the Thornycroft undertype wagon similar to one of those which competed in the War Office Trials. Its superior loadspace compared with the Foden is apparent. The Thornycroft's spring drive and lockable differential helped it in the cross-country section of the Trial.*

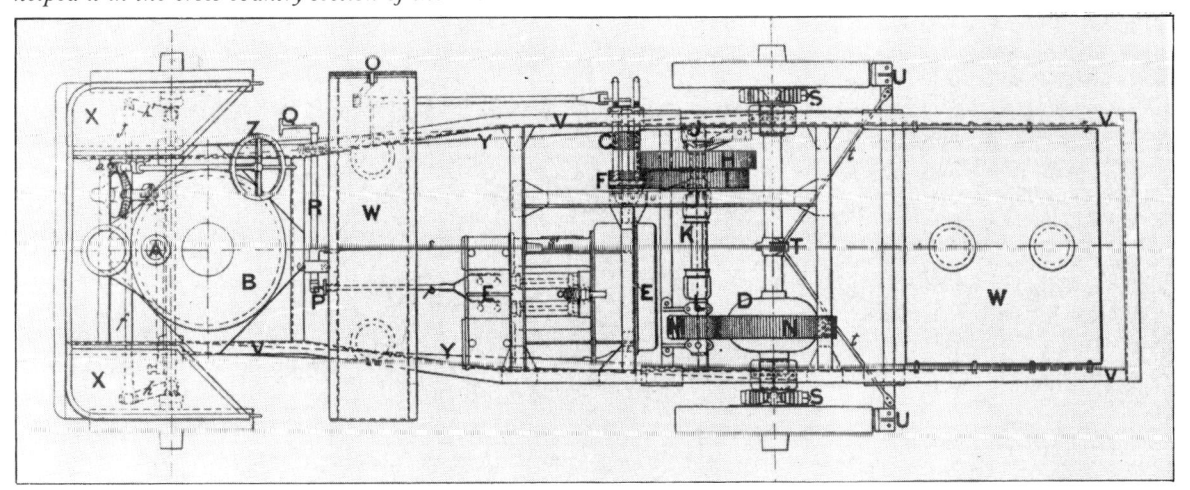

promised to be tough going. Not only were there to be three days of difficult tests on tracks and even into open country, but the final two days of the week were to be strict fuel economy tests to see how little fuel the wagons could use over a prescribed course, speed being a secondary consideration.

With their boiler cleaned, the engine and chassis newly adjusted and re-lubricated, and a bright crisp Monday morning to cheer them up, Edwin and William set off in high spirits. Over a difficult, hilly route they averaged just over 6 mph on 392 lbs of coal, while the best of their rivals was yet again the Thornycroft with 5.51 mph on 647 lbs of coal. Again the Straker used less than that, but it was slower. Edwin read the *Westminster Gazette* early the following morning with satisfaction when he came to the comment '. . . the Foden waggon was again first, for the third time in succession, covering the distance quite easily'. Reports in national newspapers were worth more than all the advertising they could ever afford to buy, and it made his years of hard experimental work seem worth every difficult moment, every drop of sweat, and every disappointment along the way. His was the satisfaction of a man whose ideas are proved right by independent examination. Similar performances followed on the Tuesday and Wednesday, and *The Daily Chronicle* reported that '. . . the steam

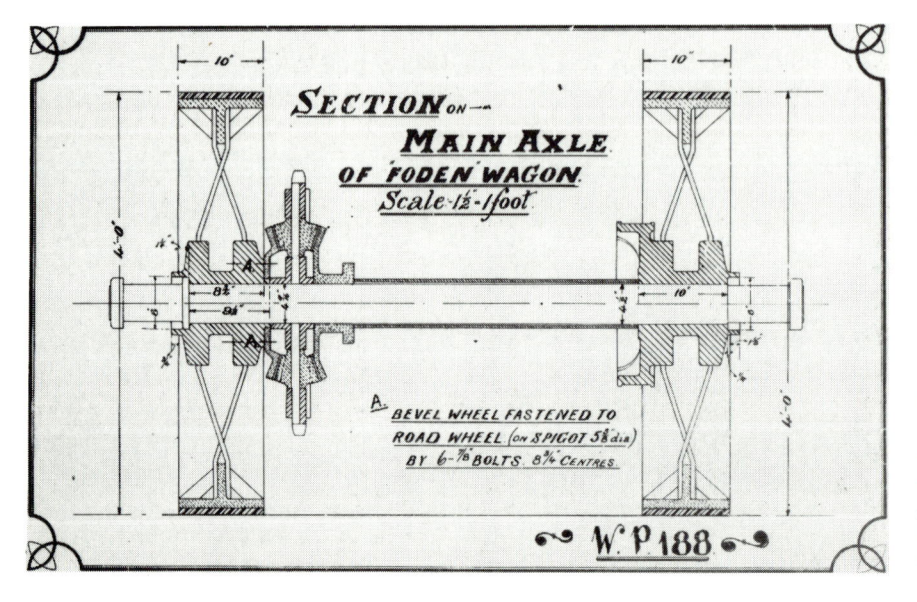

**Above left** *Edwin Foden's notes made during the early stages of the Trial compare weights and fuel loadings of 'Our Lorry' and those of the 'Thornicroft', with further notes on the first day's testing.*

**Left** *The original drive axle had outside bearings after the fashion of a railway wagon axle. The sprocket was driven by a long chain from the gears by the engine.*

## Wagons and the War Office Trials

waggon by Foden & Co, Sandbach, came in first yet again. A long hill between two and three miles in length between Odiham and Alton over a very poor by-road, tested the capabilities of the various types severely, but the Foden seemed to make light of the work'.

In fact, Wednesday December 12 was the only day on which the Foden was not fastest over the route. That day it covered a hilly circuit, on mainly good, hard surfaces, totalling 32.6 miles in 5 hours and 22 minutes, which was four minutes slower than the Thornycroft. The Straker broke down that day, and the following day too. But whereas the fastest wagon used a staggering 669 lbs of coal, the Foden used a mere 324 lbs, less than half. And that day was not scheduled as a fuel economy day, either!

The outright fuel economy runs came on Thursday and Friday December 13 and 14, which were the last two days of the road sections of the trial. Not only did Edwin and William steam their creation home in the fastest time on both days, but they used respectively 400 lbs and 402 lbs of coal on the 34-mile route, compared with 648 lbs and 667 lbs respectively for the second-place machine, which was again the Thornycroft. *The Echo* told the public in London the next morning, 'Yesterday's tests were made as to which could cover the required distance with the least fuel. The Foden was an easy winner by over 200 lbs of coal.' *The Daily Telegraph,* surprisingly enthusiastic about the whole affair considering its violently anti-road transport attitudes of a few years before, reported that, 'The Foden waggon completed the journey in under five and a half hours and used much less fuel than any other.' A little later *The Daily Express* enthused over the Foden with, 'The decided superiority of the Foden Waggon soon became manifest, day after day the sturdy little vehicle came through the trying tests streets ahead of the other waggons, and showing consistently superior economy in working, consuming far less water and coal than any of the other competitors.'

Even *The Times,* not given to extravagant words and certainly reluctant to say much about the steam road wagons in view of its staunch pro-railway editorial policy, grudgingly allowed itself to note that, 'The waggon by Messrs. Foden and Son appears to have been the most successful on

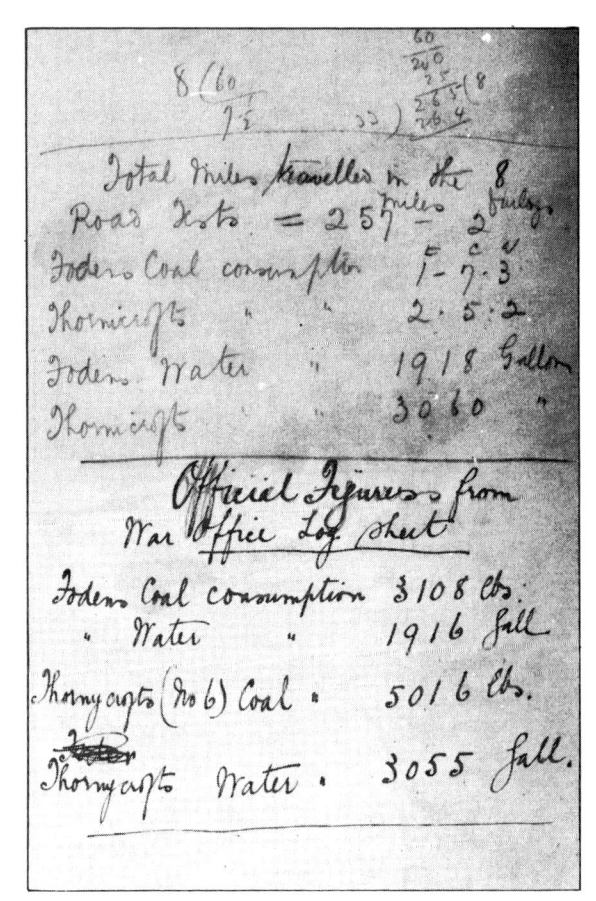

*Near the back of his Trials handbook Edwin Foden totted up his own figures for fuel and water consumption and those of his nearest rival. The official figures he inked in underneath after they had been published.*

the road trials'.

But although the Foden had indeed shown the entire entry the way home in the road trials, it took only second award in the trials as a whole. This seems very unfair looking back at results which show such a clear margin of efficiency, but the War Office committee were looking at other things too. The final stages of the trials, in the week December 16-21, involved some very difficult work in what can only be described as swamp conditions in the clay, chalk, and leaf-mould of Hampshire. The Fodens were unaccustomed to driving in these conditions, and their wheels were not really suited to it either, whereas the more local firm of Thornycroft (then based at Chiswick) had experience of just such work, had a lockable differential and had wheels to suit as well. Furthermore the Foden, being an overtype, had a relatively short load-carrying body space compared with the vertical boiler undertypes, and that brought it bad marks in the loading trials. Consequently Edwin and William went home with the £250 second prize, but they were quite happy with that. They knew that commercial operators placed economy of operation far, far ahead of any other characteristic, and they had proved beyond question that their steam wagon was far superior to anything else available in Britain at that time, indeed anywhere in the world.

As soon as the trials were over they cleaned their boiler again, did some routine maintenance and greasing, and set off to steam the 160 miles back to Cheshire in the depths of winter with their successful wagon. The gentlemen of the War Office committee had decided to buy the wagon, but first Edwin wanted to make some minor modifications before their lordships took the machine for extended evaluation. The Fodens arrived back at the factory very late on Christmas Eve 1901 to a truly tumultuous welcome, not only from their excited employees but also, it seemed, from a large proportion of the populace of Cheshire, who had been following the progress of their local engineers day by day in the newspapers. To the workforce at Elworth it was quite a Christmas present. Although they were artisans and not economists, they knew that the success of the wagon they had helped to build could only mean prosperity for the company, and that meant full bellies for them and their families in the future.

As Edwin and William steamed their travel-stained but sweetly running engine into Sandbach marketplace and down the road towards Elworth — having covered nearly 600 miles in a little over three weeks — they knew that things would never be quite the same again. The dreams that Edwin had described during his nostalgic chats with George Hancock about the great steam carriage days, had finally become reality. The War Office committee issued a statement after the trials had been completed and studied, which included the following passage: 'Steam lorries are good and serviceable machines suitable for present supply, and likely to be of great advantage to the transport service in country where fuel and water in sufficient quantity is available. The Trials have shown that self-propelled lorries can transport at least five tons of stores at a consistent 6 miles in the hour over considerable distances on hilly roads in winter conditions.' It went on to explain how many horses would be needed to do the same amount of work, and how much longer it would take.

The point was made, and almost every newspaper in the country reported on these exciting pronouncements at length. To Edwin Foden, then just over 60 years of age, it meant that his life's ambition was close at hand. Road freight transport was not only feasible, it was going to be in big demand, and very soon. Whereas he and William had set out from Elworth before the event with little more in their minds than getting their wagon down to Hampshire and putting it through the trials programme to the best of the abilities of both men and machine, they returned with greater things on their minds. They would have to begin major reorganisation and expansion at the works if the natural sequel to the trials success was to be exploited fully. The £250 prize cheque tucked safely in Edwin's inside pocket would not go far towards financing that expansion. There was work to do.

# Chapter 6

# Into the golden era

Even before Edwin and young William steamed their victorious and already famous little three-tonner back to Elworth on that joyous Christmas Eve in 1901, Edwin was acutely aware that the success of their venture down to Hampshire could not be properly exploited with the facilities and capital then available. That Christmas was a thoughtful one for him. While the traditional family celebrations in the Foden home took place, Edwin's mind was not entirely occupied with food, drink, or Watch-night services. Not that he neglected his family or his duties — he was too staunch a Methodist and too closely tied to his family to think of such a breach of good manners — but slowly he formed the outline of a plan that would take his company to the very top of the industrial tree in Britain, and make the name Foden synonymous with reliability and strength in the field of transport.

The old company structure was woefully inadequate for the new tasks, and he had seen too many excellent engineering concerns go under simply because their management structure fell short of the excellence of their engineering. So a new company was formed, named simply Fodens Ltd. As might have been expected — and who would argue anyway? — Edwin Foden was managing director of the new board, while he invited a business colleague, Mr Cecil H. Brunner, to be chairman. Mr Brunner was chairman of the nearby firm of Brunner Mond Ltd (which is now part of ICI), and had great experience in financial matters. The Stubbs brothers, John and Reuben, who had been such good if unorthodox customers at the Elworth Foundry over the preceding 20 years, were invited to join the board, and the Stubbs family has had a direct involvement with Fodens ever since. Both of Edwin's sons, William and E.R., joined the board, and brother-in-law S. P. Twemlow continued as company secretary — a post he had held since 1886. Together they set about raising new capital. The board of the new company had an excellent blend of engineering, commercial and financial expertise, and few small firms — which Fodens still was — can have had such a highly able collection of directors with which to establish its reputation and fortune in the world.

The nominal capital of the new company was £100,000, although of that only £58,552 10s was subscribed. The record books show that at the inception of the new firm its buildings and land holdings were valued at £12,042, while plant, machinery and tools were valued at £24,977. These figures are taken from the first balance sheet, so it is probable that they show the value after the intial re-equipment of the old Elworth Foundry. During that first momentous year, 1902, the old foundry building was rebuilt and enlarged, and the first of a series of additional machine shops and workshops was added to accommodate the new steam wagons that were being ordered and sold in increasing quantities. Across the road from the foundry itself, a new office building (still occupied today) was erected, using a very high quality semi glazed brick for the facings. This brick was subsequently used on all the offices and houses built by the company, as we shall see in another chapter. The street in which the extended factory was built, Elworth

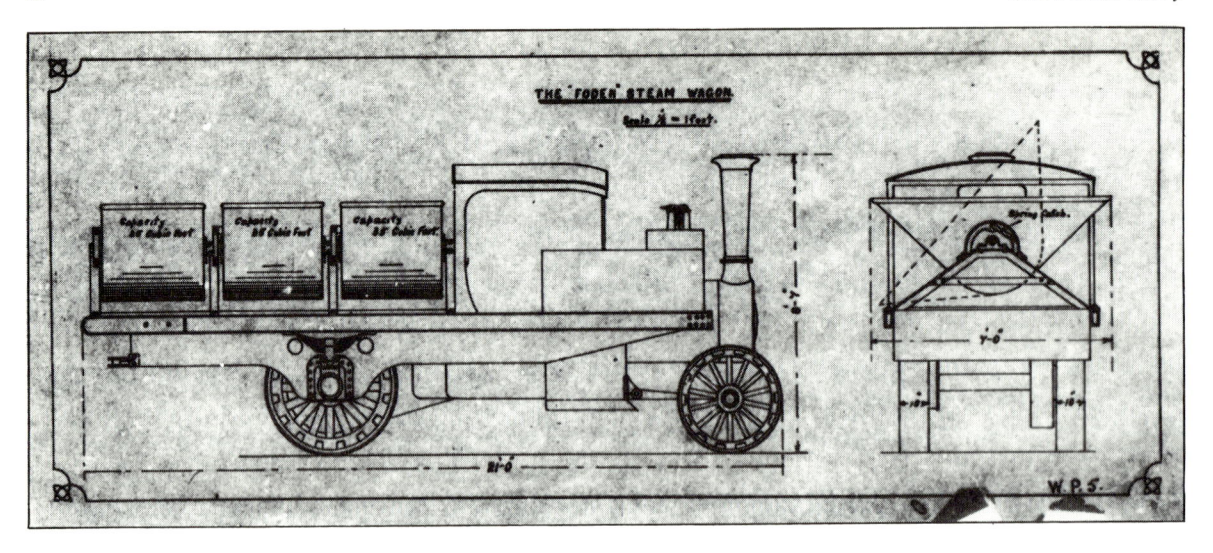

Road, still forms the main thoroughfare of the Foden works, and over the years it has seen many triumphs and tragedies.

Such was the extent of the original expansion that 1902 was almost over before the entire works was fully operational. As it happened, this tied in nicely with the development of the wagons themselves because, successful though the War Office machine had been, Edwin was determined to improve it, and during those 12 months from Christmas 1901 a great many changes were made to the design, mostly aimed at improving reliability, simplifying maintenance and repair, and improving economy still further. This particular programme is discussed in the chapter on wagon development.

The worth of the new board was proved over and over again during that year, with the engineering members concentrating on their developments in the factory, while the commercial

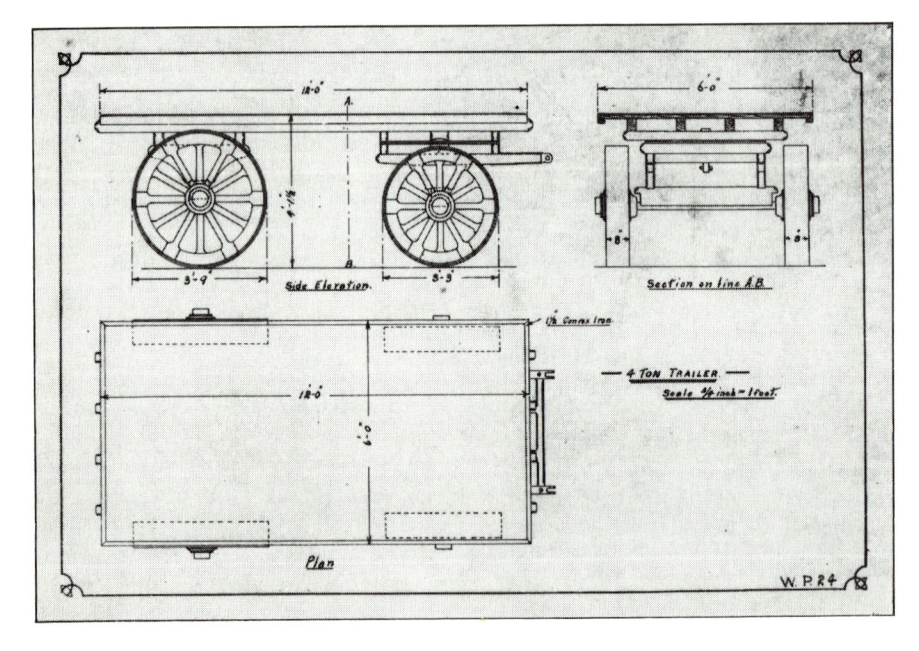

**Above** *Layout drawing of a side skip-tipper, dated 1902, shows the outside frame arrangement used on the first wagons.*

**Left** *Trailers were offered for a number of years as an accessory to the wagons. A four-ton design of 1904.*

**Above right** *The first major change in design on the three-tonner wagon was the location of the main frame rails inside the rear wheels instead of outside as on the originals. On the early models, braking consisted of a friction block on the iron wheel rims. These wagons were built prior to the new Heavy Motor Car Order.*

Into the golden era

men got on with setting up a proper sales and repair facility to handle the new products.

Just how successful the new team proved can be seen from the fact that, as soon as the reorganisation was complete the company was showing a profit. The balance sheet for 1903-4 shows a healthy gross profit of £11,353 7s 7d and, after tax, dividends to shareholders and sundry other items were deducted, there was still a net profit of £5,650 4s 2½d to be ploughed back into the reserve. This was a very satisfactory result when you consider that a steam wagon complete with body could be bought for under £500. Profits increased steadily during the following years as the formation costs of the company were absorbed and trade boomed, until by 1907 the gross profit was £19,797 with a net figure of £7,332. By then the shareholders were enjoying regular dividends of 12 to 14 per cent, and the annual reports from that period indicate a general air of prosperity almost bordering on euphoria. On June 23 1907 John Stubbs, who had for so long been a confident, customer and friend of Edwin Foden, died and he was sorely missed. His was the commercial skill in the outside world that had led Fodens Ltd to prosperity. Samuel Poole Twemlow, who had by then been company secretary for 20 years, was elected to the board — the first of a long line of Twemlows at director level within the company.

A curious sideline product persisted for some years after Fodens became a modern industrial concern. Way back in the 1870s and 1880s one of their most successful machines had been the portable steam threshing-drum, and, in fact, one of their numerous tasks had been that of agricultural contractors specialising in corn threshing. This sideline was retained after the formation of the new company, largely because Edwin Foden did not want to lose contact with his agricultural friends — who were good potential wagon customers — nor did he want to deprive the local farmers of the service they had enjoyed for so long. This was no gesture of philanthropy because, right up until 1909 when the threshing work was eventually discontinued, it made a modest profit, contributing something like £300-£400 a year to the overall gross profit.

By the time a decade of Fodens Ltd was completed gross yearly profits had risen to £59,727. Satisfactory though this was, there were always new problems. Edwin Foden died on August 31 1911 and, although this event must have been expected for quite some time, his passing hit not only the company but the entire Elworth community as a major disaster. To them Edwin Foden was more than an employer; he was for most practical purposes their guardian and benefactor. For almost as long as the majority of the employees and their dependents could remember, Elworth Foundry and Edwin Foden had been synonymous with a progressive community, a steadily rising standard of living and a growing pride in the knowledge that their small Cheshire town was famous for its

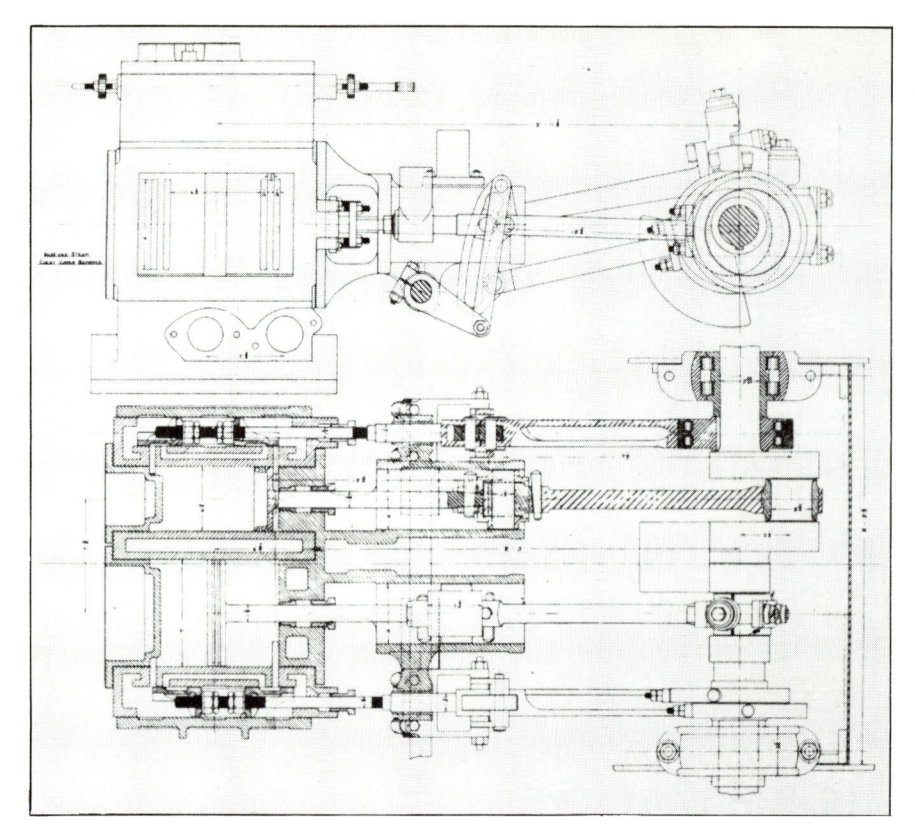

**Left** *Foden's steam-wagon success revolved around the compound engine. Particular features are the 'double-D' valve arrangement and the roller-bearing crank.*

**Below right** *Typical application of the five-tonner, moving furniture in a van and trailer. Note the 'one-leg-out' driving position employed on the early overtypes. The year is about 1907.*

**Below far right** *During World War I large numbers of heavy shell casings were cast and machined by Fodens Ltd. Not one shell was ever rejected by the War Office inspectors. E. R. Foden can be seen third from the left.*

engineering products. They believed that when Edwin Foden died the company would also lose his strong influence and leadership. In the literal sense, they were of course quite right. And somehow the two 'boys', Willy and E.R., then aged 43 and 41 respectively, did not measure up to Elworth's memory of what their father had been. Had not Edwin, after all, managed the affairs of the thriving company almost to his deathbed? The day of Edwin's funeral was a sadder occasion for Elworth than the passing of Queen Victoria 10 years earlier. It wasn't that they did not love their queen, but Edwin Foden was different. He knew them all by name, took an interest in their affairs, made sure that everyone had at least the minimum necessities of life, even if it required subtle methods to avoid the appearance of charity. When the Foden Motor Works Band led the *cortège*, playing Edwin's favourite Methodist hymns, there was scarcely a dry eye in the whole of that part of Cheshire. Many a sage predicted that things would never be the same again. There could never be another Edwin Foden.

They were quite right, of course: there would never be anybody quite like Edwin, and nobody was more aware of the fact than his two sons. They had been quite content to leave the helm of the company to their father; after all it had been his engineering and commercial brilliance that had created it all. But they had not sat idly by watching the family business flourish. Both of them knew a great deal about steam, had enormous engineering expertise, had acquired a keen commercial sense from Cecil Brunner and John Stubbs, and above all were trusted by the workforce. It was quite common for both Willy and E.R. to be found in overalls on the floor of the boiler shop or the erecting shop, working on a new design or development with the foreman and leading fitters, translating ideas into practice with the aid of lots of chalk sketches on the floor and on pieces of steel plate. Consequently, when the mourning for their father was over, the Foden brothers tackled the task of speeding the company

*Into the golden era*

on its way to success in a manner that surprised even their most fervent supporters.

The immediate task they set themselves was to invest heavily in both buildings and machinery, so as to expand an operation that they now knew was second to none in Britain, indeed in Europe, as far as the manufacture of heavy-duty freight-carrying vehicles was concerned. Not only was the Foden the most successful steam wagon, but the internal-combustion-engined machines had made little progress beyond the light truck stage, and certainly could not compete in ruggedness and durability with the highly developed steamers in the 5-ton load class. By the end of 1911, almost £8,000 worth of new machinery was installed, and £2,400 was spent on extra buildings. Over the next four years an astonishing total of £38,000 was spent on additional machines and tools, while a further £35,000 went into new buildings, including many new houses for company employees. It was during this period, in fact not long after the passing of Edwin, that Francis Poole joined the board, to remain a central figure there for many many years.

There were those who questioned this seemingly extravagant expenditure, not least some of the shareholders who wanted to know why spending in 1913, for example, on machinery and buildings alone, should be more than the gross profits of a couple of years previously. But Willy and E.R., ably monitored by the chairman, knew exactly what they were doing. The shareholders were effectively silenced by the distribution of £61,480 in the form of special bonus shares on a one-for-one basis, and the board got on with what it felt it had to do. The gross profit figures ran at approximately double those of three years previously for quite a while.

For example, the 1912 figure of £59,727 was just over twice the 1909 figure. That continued to 1914 when the gross profit of £86,926 was almost double the figure when the two brothers took over in 1911.

It had become clear to most thinking men, and certainly to Willy and E.R., that war in Europe was inevitable before long and, much as the prospect appalled them, they realised that if Britain were to fight a land war successfully, transport would be needed for supplies, munitions, heavy guns, and all the paraphernalia of war. So, while a rapidly increasing number of Foden wagons were built and sold to civilian users, both at home and overseas, the Elworth factory began to turn out slightly modified machines for the War Office. Although the military authorities had purchased occasional wagons after the famous trials, they were strangely reluctant to employ mechanised transport in any of their training manoeuvres, due largely to the strength of the long-established cavalry regiments, who insisted that their brand of warfare was still the most effective. However, War Office attitudes had changed somewhat by 1913, when a trickle of orders for mechanised transport began, Fodens' first being for 20 5-tonners.

At the same time, the War Office initiated its subsidy scheme, under which approved types of truck could be bought by hauliers at a state-supported discount — usually about £50 — and they would receive an annual maintenance grant in return for making the vehicles instantly available to the authorities in the event of mobilisation. Leyland and Thornycroft were the principal beneficiaries under this scheme. However, thoughts were already turning away from steam towards internal-combustion-engined trucks in some quarters, and in the event there were many thousands of petrol-engined trucks employed during the war years, as against something like 800 steamers. Although steam vehicles proved to be much more durable under war conditions, the long process of starting a steamer from cold was not always compatible with the logistics of warfare. On the outbreak of war in August 1914, large numbers of wagons

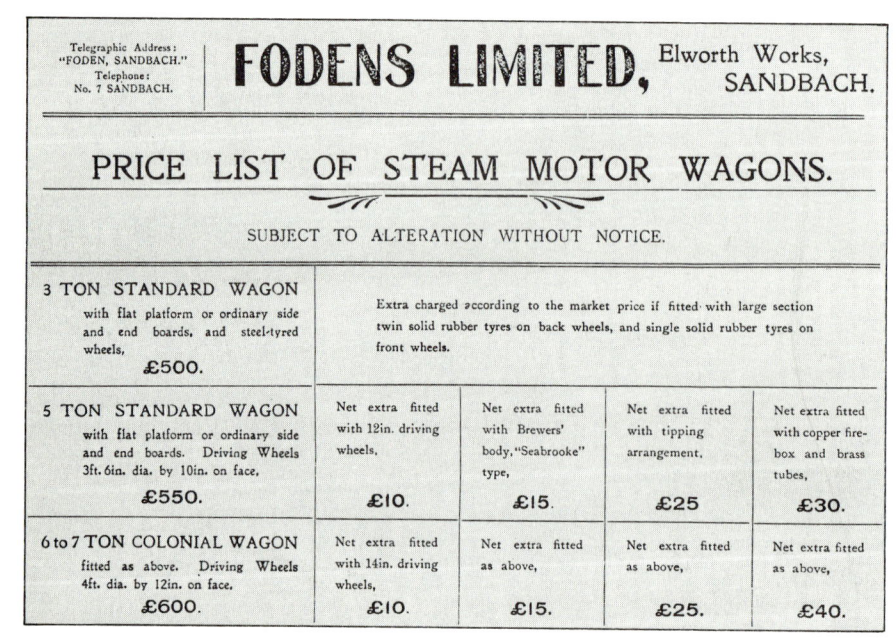

Telegraphic Address: "FODEN, SANDBACH."
Telephone: No. 7 SANDBACH.

# FODENS LIMITED, Elworth Works, SANDBACH.

## PRICE LIST OF STEAM MOTOR WAGONS.

SUBJECT TO ALTERATION WITHOUT NOTICE.

| 3 TON STANDARD WAGON with flat platform or ordinary side and end boards, and steel-tyred wheels. £500. | Extra charged according to the market price if fitted with large section twin solid rubber tyres on back wheels, and single solid rubber tyres on front wheels. ||||
|---|---|---|---|---|
| 5 TON STANDARD WAGON with flat platform or ordinary side and end boards. Driving Wheels 3ft. 6in. dia. by 10in. on face, £550. | Net extra fitted with 12in. driving wheels, £10. | Net extra fitted with Brewers' body, "Seabrooke" type, £15. | Net extra fitted with tipping arrangement, £25. | Net extra fitted with copper fire-box and brass tubes, £30. |
| 6 to 7 TON COLONIAL WAGON fitted as above. Driving Wheels 4ft. dia. by 12in. on face. £600. | Net extra fitted with 14in. driving wheels, £10. | Net extra fitted as above, £15. | Net extra fitted as above, £25. | Net extra fitted as above, £40. |

*Into the golden era*

**Far left** *A dozen five-tonners prepare to steam off to war in 1915. This picture shows clearly the curious 'one-leg-out' driving position.*

**Left** *The financial fortunes of the company ebbed and flowed over the years. This graph shows the pre-tax profits annually from the formation of Fodens Ltd to the outbreak of World War 2.*

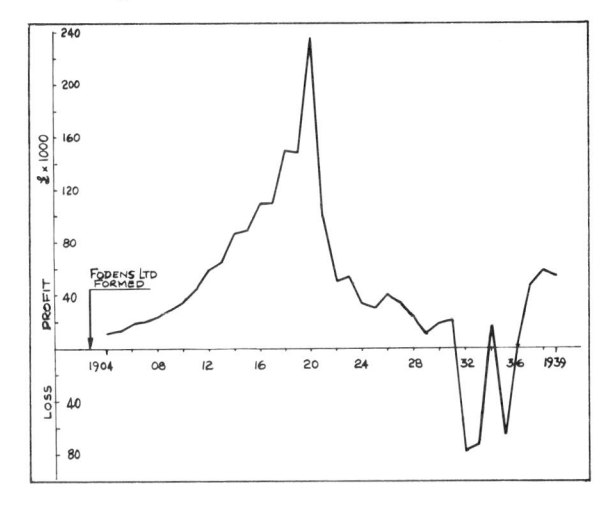

were commandeered by the War Office, on the basis of the price as new less 15 per cent for each year of use.

The early days of the war were treated very much as a big adventure by most Britons, and there were few who imagined that it would last more than a couple of months. Large numbers of men went off to join the forces, and the depleted ranks at many factories, Fodens included, meant a slow-down in industrial production. Not until a great amount of labour reorganisation, involving increased use of female labour, had taken place did the situation return nearer to normal. Consequently, we find that Fodens' gross profits for 1915 were only about £1,300 up on 1914, at £88,269. That year one of the biggest orders yet for Fodens Ltd came from Pickfords, for 50 steam wagons. It is worth noting that even then, Pickfords had already been established for 140 years as carriers, and the famous name did no harm at all to Fodens' reputation.

But, as the war ground on into 1916, it became painfully obvious, even to the most optimistic of Britons, that this jolly little war was going to turn into one of the bloodiest conflicts the world had ever seen. The mood in Britain changed. It took a long time for the message to sink in that everybody would have to put their utmost into the war effort if they were to survive, but gradually more and more of the workforce took up multi-shift working.

At Fodens it was no different. Wagons continued to be produced, but the little Cheshire factory's high skills in foundry work, together with its machine shops, made it an ideal location for the manufacture of heavy munitions, mainly 9.6-inch shells for use in naval and coastal guns. Even the bandsmen, who had previously been pure musicians, went to work in the factory, and most men worked at least a 12-hour day. For most of the second half of the war, work at Elworth went on round the clock, and it speaks volumes for the quality of production that, of the many thousands of shells made, not one was ever rejected by the Ordnance inspectors.

**Far left** *The 1914 price list looks like a bargain sale. The £25 for body tipping gear would scarcely buy a coat of paint for the body these days.*

**Left** *A few D-type tractors were built for showmen's use. With solid rubber tyres they were capable of about 25 mph, though the legal maximum was 12 mph.*

54                                                                                                          The Foden Story

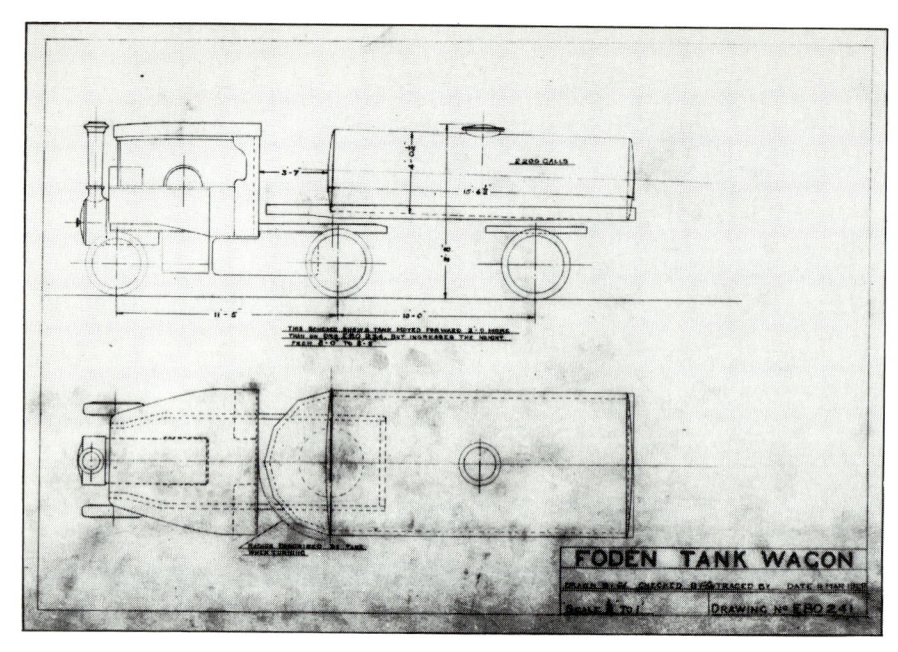

**Left, below left and below far right** *Innumerable versions of the highly successful C-type wagon were built, of which these are just a few.*

**Below** *Trailers continued to be part of the activity at Elworth for many years. This three-tonner was built about 1922.*

All this frenzied activity was reflected in the financial results, and the 1916, 1917 and 1918 gross profits were respectively £109,501, £111,066 and £149,610. Clearly war was a profitable business for industry, if not for the nation as a whole, but Fodens was not the only company to make a lot of money at that time. Leyland, AEC, Thornycroft, and a number of other vehicle builders made tremendous strides during those four war years, but whereas most of them found the business climate distinctly chilly after the Armistice in November 1918, with huge numbers

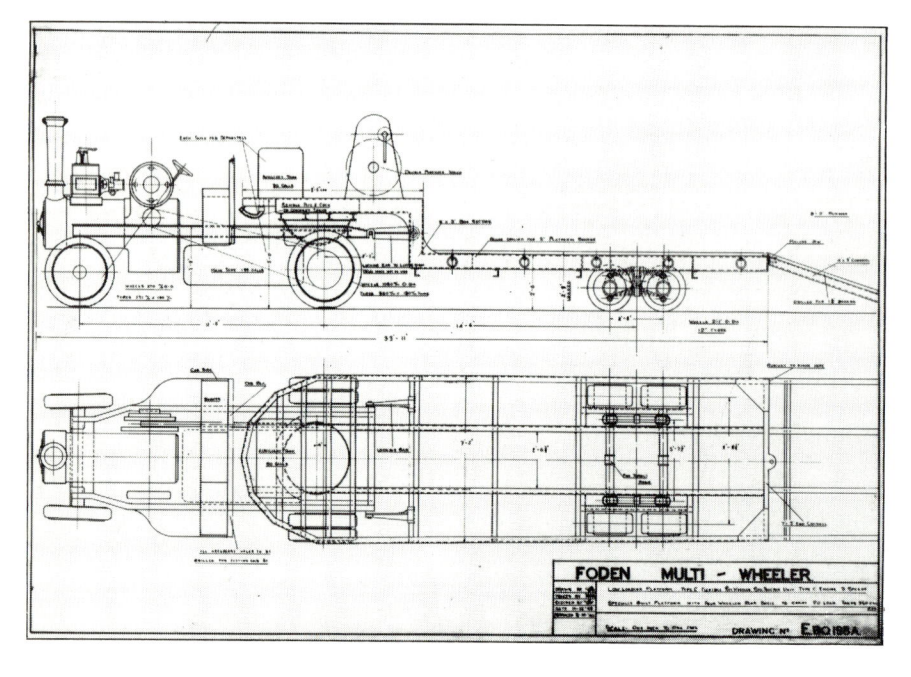

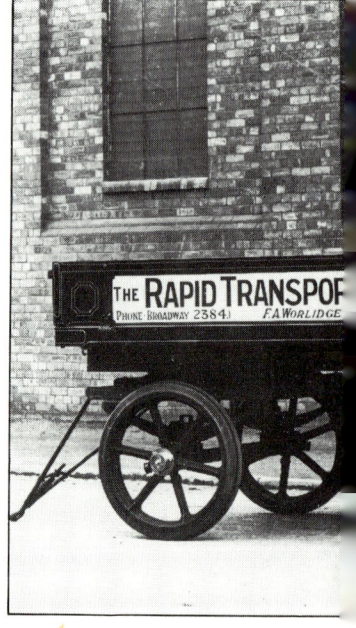

*Into the golden era*

of ex-military lorries around blocking the sales of new machines, Fodens Ltd managed, at least for a short time, to go from strength to strength.

The main reasons for Fodens' prosperity lay in the petrol supply situation: more than one authority, although they ought to have known better, loudly estimated that the world's entire petrol supplies could last only another three years at the current rate of consumption. Heavier taxes were placed on petrol, particularly in 1919 and 1920, and at one time its price was 4s 3½d a gallon. As the average five-ton lorry would do only about three or four miles to a gallon at that time, many hauliers and own-account users of transport invested in steam wagons. They cost more to buy — say £900 on average against half that for a petrol machine — but the running costs were more predictable and the life expectancy and reliability of the steamer were still better than in most petrol-engined lorries.

The shareholders, who had received a further bonus share issue on a three-for-two basis just as the war ended, were clearly in an anxious mood at the next AGM in the summer of 1919, and many questions were asked about the future. Francis Poole was chairman by that time, as Cecil Brunner had retired, and Willy, E.R. and Sam Twemlow were joint managing directors. Fodens' market position seemed sound enough, despite grave doubts about the country's economic future in general, and this message was passed on to the shareholders. They were not disappointed, for the 1919 results showed a gross profit of £148,782, resulting in a record 18 per cent dividend.

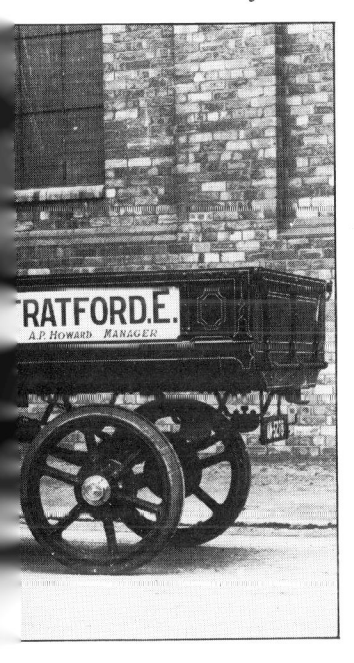

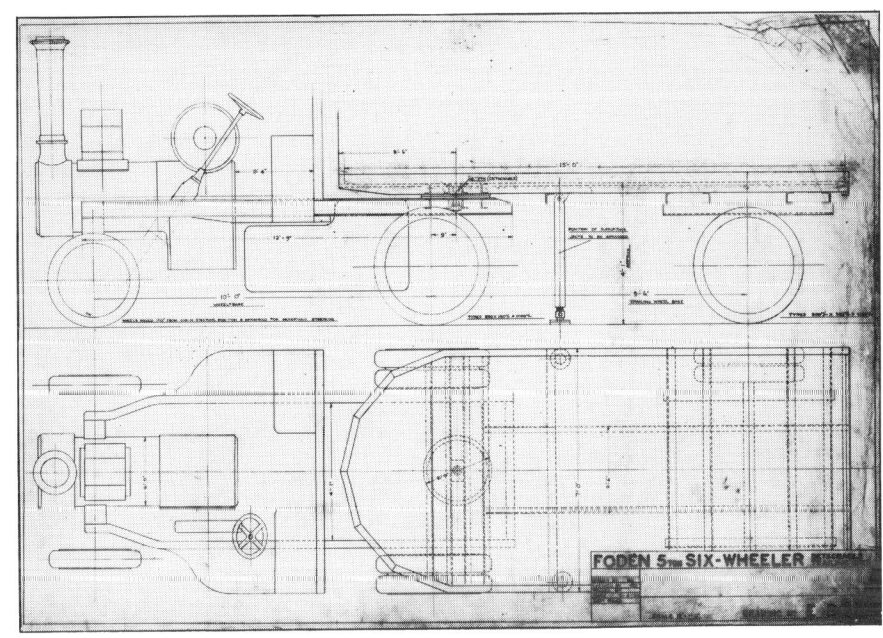

The winter of that year saw the first signs of trouble, though it scarcely affected Fodens. Men who had fought for long years at the front, and seen hundreds of their comrades die alongside them, were dissatisfied — no doubt justifiably — with the poor working conditions, the meagre wages, and the lack of jobs in many parts of the country. This was not, they argued, the land fit for heroes to live in, which they had been told about so often while they were fighting. The first big strike was by the coal-miners, but fortunately it was short-lived. Rather more serious was a strike of foundry moulders in the winter of 1919-20 which began in the Black Country and lasted for 19 weeks. Within a week or two of its beginning the strike had reached national proportions, but such was the loyalty at Fodens, and so strong the relationship between management and workers, that only a small proportion of the Foden moulders joined the strike. This was indeed fortunate for the firm, because the production of their steam wagon was very dependent on the availability of top-quality castings for engine parts, transmission components, and above all the road wheels.

Despite the uneasy industrial mood, such was the volume of business, and so well did the company manage to organise its affairs, that a record trading year resulted, with gross profits of £234,980. This meant a further 18 per cent dividend for the shareholders, but it was not only the investors who reaped the rewards of that unprecedented trading result. The company had already established a good reputation for looking after its workers, and had built a large number of houses just before the war. That winter of 1920-21 saw a further 26 cottages built for employees; in fact Fodens Ltd even bought a brickworks to support their building activities in Elworth. In that year too a superannuation fund appears for the first time in the company records, with £10,000 as the basic investment. But all was not bright on the horizon. At the AGM there was talk of 'the greatest slump the world has ever known', and on a reduced trading volume the profits shrank to £99,736 — poor by the previous year's standard but still a lot healthier than many other heavy-vehicle manufacturers could achieve. The clouds did not deter the board from its plans, however. Their engineers had developed a new, more efficient 6-ton steam wagon called the C-type, and over £13,000 in new machinery and tools was invested as part of the firm's future life-blood.

By 1922 the C-type was in production and selling reasonably well in comparison with competitors' machines, but money was now in extremely short supply, strikes in one industry or another were a constant source of trouble, and export markets had all but dried up. At the 1922 AGM Francis Poole told the shareholders, 'The cost of our production is still far too high to keep us in a competitive position, and wages will have to come lower if we are to keep the works going.' Nevertheless, trading was still profitable, with a gross figure of £51,469, and the following year it actually increased a little to £54,091, thanks largely to the success of the C-type. A further model was added, the 10/12-ton six-wheeler, to cater for the high payload demands from some sectors of industry, and the threatened wage cut-back did not become a reality — at least not that year. Nevertheless, the dividend was down to five per cent, and there were many dark mutterings at the AGM about poor trading to come, excessive costs, and the extreme scarcity of orders.

A reflection on the state of business was that under the 'sundry debtors' heading on the 1924 balance sheet was a figure of over £89,000 for deferred payment sales — something that Fodens had never before found necessary. However, with the depression taking a firmer grip, this was the only way that customers could afford to buy the expensive steam wagons, and any sales were better than none at all. In 1924, at his last AGM, Mr Willy Foden explained that there was both good and bad news. Profits were down to £33,071, the dividend was only five per cent, and there were only three unfulfilled orders in the book. 'But we are satisfied that we are doing better than anyone else in the industry,' he told the shareholders. 'We are however severely hampered by the very large numbers of second-hand motor vehicles currently on the market'. Many of these were, of course, reconditioned wartime trucks, being sold at very low prices indeed, and naturally many hauliers chose to

*A tracked version of the normal wagon was built in 1924 as an off-road machine. It was not a success, mainly due to steering difficulties.*

*Into the golden era*

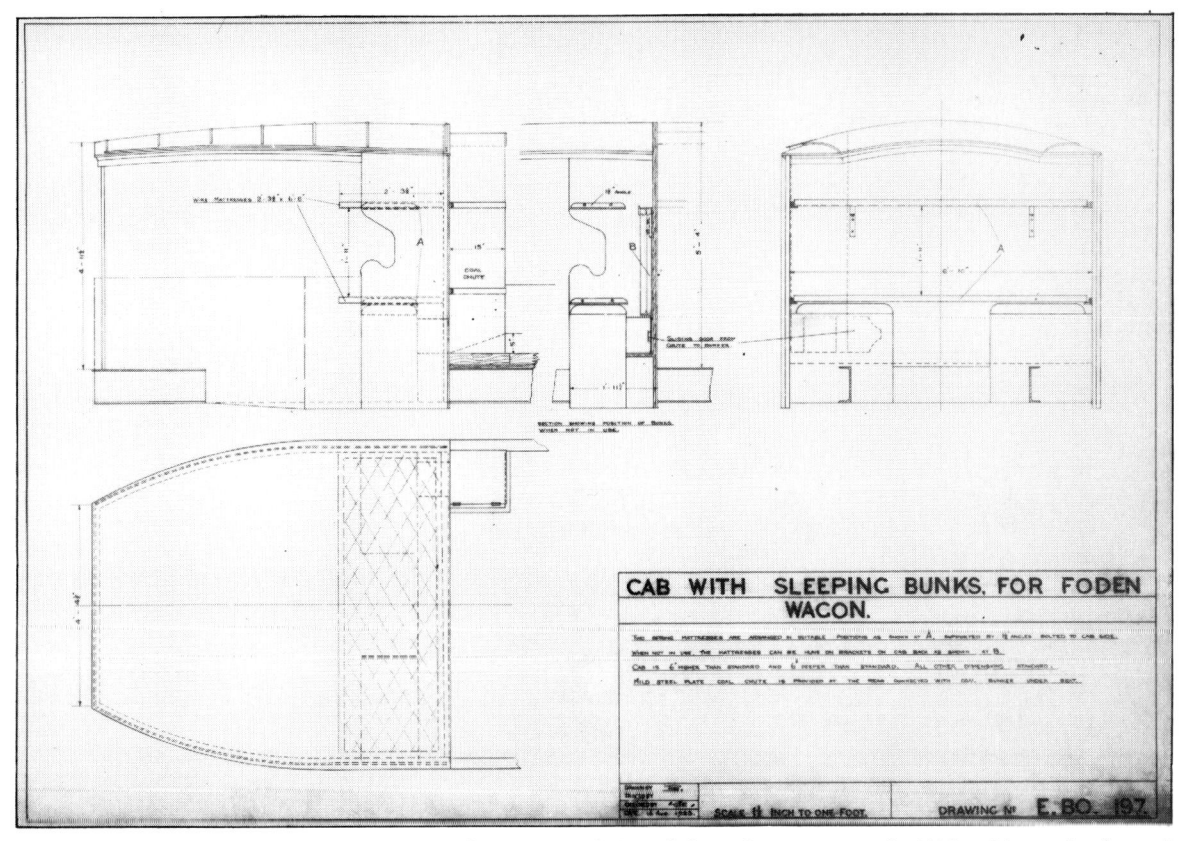

*Sleeper cabs are not a 1970's phenomenon: this one was designed for a C-type wagon in 1928, with two bunks each 6 ft 8 ins long and 2 ft 3 ins wide.*

keep going on the cheapest available equipment; who could blame them? The good news was that King George V had ordered a 5-ton Foden steam wagon for use at Balmoral — there was already a Foden timber-wagon at Sandringham — and a new 'undertype' wagon was being developed and tested that would, it was hoped, be more efficient and therefore more saleable, than the long-established 'overtype'.

But by this time the strain of the constant struggle to keep things going profitably was beginning to take its toll. Willy was now 55 years old and he felt that, having seen the company through the birth of the steam-wagon age (with his father) and a great world war, it was time he settled down to something a bit less demanding. In November of 1924, Willy and his wife, his two sons Ted and Reg Foden, along with Reg's wife and father-in-law, sailed for Australia and set up as sheep farmers on an 8,760-acre station near Newcastle, NSW.

The factory staff and workers were dismayed. It had taken them a long time to get used to the absence of Edwin Foden, and, though none of them questioned Willy's reasons for emigrating, their fears for the future of the company and themselves were deep indeed. There was some consolation in the fact that the youngest son, E. R. Foden, was to remain as the sole Foden on the board, but although he was well liked, and highly respected for his engineering skill, the future looked bleak. In fact the employees were quite justified in their concern. Never again would the happy Foden ship be quite as it was in the golden days before World War 1, and in the less golden but very prosperous years immediately after that conflict. Now that Mr. Willy was gone, another era had ended.

# Chapter 7

# A symphony in steam

Present-day truck-men complain bitterly that the development of the vehicles from which they earn their living is far too closely controlled by legislation on all levels and by a whole host of other factors outside the scope of engineering design. It may come as some surprise, if not much consolation, to learn that Edwin Foden was making exactly the same complaint at the very wakening years of this century when his first steam wagons were being produced. The 'Emancipation Act' obliged the manufacturer to keep the weight of his vehicle below three tons unladen, because only by so doing could an economically viable speed — 12 mph — be achieved legally.

When you consider what went into the first steam wagons, with only a very short development time behind them, the fact that they could be kept down to anything like that three-ton limit seems little short of miraculous. Boiler, firebox, engine, transmission, axles, frame, wheels, body, canopy — all were built substantially so that they would 'live' under the hardest work conditions, and the amount of detail engineering that must have been put into each and every part in order to make it sufficiently strong but as light as possible, I leave to your imagination. To put this weight problem in context, it is only a modest-sized truck in this modern age that tips the scale at three tons — something like a Leyland Terrier or a Bedford KD, and it seems inconceivable that the Victorian engineers could get the weight of a steam wagon down to a similar level. Of course, by the time the boiler and water tank were filled, and something like a ton of coal was aboard, the weight would be nearer five tons, but there was still plenty of capacity for a three-ton or more payload, and of course an additional load went on the trailer as often as not. But the key factor was the three-ton unladen weight, and once that was achieved the wagon was a viable proposition. In fact a number of would-be wagon builders went out of business overnight simply because they could not meet the weight, and few users wanted a wagon restricted to 4 mph when they could buy one that would do 12 mph, at least theoretically.

Once the basic design was developed and proven, as it so successfully was in the War Office Trials, there were plenty of enquiries from industrial users. Probably the first firm to put the new Foden wagon into commercial service was far from Elworth — though, perhaps significantly, not far from the trials area — and that was W. H. Brakspear & Sons, brewers from Henley-on-Thames. Their machine had the very early type of fabricated frame arrangement with the outer rear-axle bearings supported on angle-iron brackets riveted to the main side-frame members. Very shortly after this a lighter, stronger design was evolved in which the bearings were carried in a specially shaped deep side-member formed in steel plate. This member was still outside the wheels, which at that time were 4 feet in diameter at the rear and 3 feet at the front.

The provisions of the Heavy Motor Car Order of 1904 made history. Not only was it one of the very few times in the entire history of transport legislation that the road vehicle builders got more or less what they wanted, but it also established

## A symphony in steam

the heavy commercial vehicle industry in Britain as a viable competitor to other forms of transport for the very first time. Despite its official title, the order did not come into force until January 1905. It was principally concerned with overall weights, axle weights, wheel sizes, brakes, registration marks, lighting and suspension. A 'heavy motor car' was permitted to weigh up to five tons unladen and there was an additional class for older machines up to seven tons: a trailer could weigh up to one and a half tons. But it was the laden limit that made the most interesting reading, for up to eight tons gross could be carried by any one axle, and up to 12 tons on the whole wagon. A further eight tons could be accounted for by a trailer, so that for the first time a 20-ton gross combination could be operated legally, and that meant a payload of something like 14 tons in the case of a thoroughly modern steam wagon. The speed limit was rather complex but broadly speaking 5 mph was the maximum if a trailer was being towed, and 8 mph without. If pneumatic or rubber tyres were fitted, then the solo speed could only be 12 mph if the heaviest axle weight did not exceed six tons. Just how the village constable was meant to work all that out as the wagons steamed through is not recorded but, as we shall see in another chapter, some of them did, with good effect.

Another far-reaching aspect of the new regulations was that concerning wheel sizes. A very complicated formula was supposed to

*Precise wheel dimensions related to weights were laid down by the Heavy Motor Car Order 1904. These tables refer to the wheel sizes used by Fodens.*

govern this, but to help those not mathematically inclined, tables of permissible sizes and weight ratings were published.

There was a great deal more detail, though by no means as much as current Construction and Use regulations contain. The whole piece of legislation was written in perfectly straight-forward English: the tangled official language with which we struggle today had clearly not been invented in 1904. The reason for allocating so much space to this legislation is that it was undoubtedly the most important document in the early development both of the steam wagon and indeed of the entire road haulage industry.

When the details of the Heavy Motor Car Order were published in the autumn of 1904, wagon production was already in full swing at Elworth, some 30 machines having already been completed. The Fodens had a very shrewd idea what the new regulations would contain and they had already worked out roughly what their plans would be. Early in 1905 they announced their new 'five tonner' which took full advantage of all the new regulations, although retaining the basic design of the War Office Trials machine and its successors. The wheels were smaller, the layout permitted even loading right up to the maximum limit, the frame sides were inboard with wheels out at maximum width, the compound engine was slightly larger than on the earlier models and developed about 20 hp. Two-speed gearing was continued, with the same arrangement of spur gears and a long driving-chain to the rear axle, and the driver still sat on the left to drive the wagon. The load platform was over two feet longer than before, now measuring 11ft 6ins by 6ft 6ins, and the maximum designed speed was about 15 mph, though of course the legal limit was less.

Such was the success of Edwin Foden's wagons that there were many imitators, if not of the whole concept then of the detail design. One such, considered Edwin, was the Hampshire firm of Wallis & Steevens, who had built steam traction-engines from the early agricultural engine days, and who had switched to wagons in

*The five-tonner model took full advantage of the provisions of the Heavy Motor Car Order and brought real freight moving ability to the new road haulage industry.*

*A symphony in steam*

1905 when a prototype was built, which was developed into a production machine in 1906. Wallis & Steevens were never big-volume producers of wagons — they built only 127 machines spread over 18 years, whereas Fodens built something like 7,000 wagons in all — but, when the Wallis & Steevens wagon first appeared at a local show near Basingstoke, there were certain features about it that Edwin Foden felt infringed his design. The main bone of contention was the way in which the frame rails were swept in at the front to fit snugly alongside the boiler, and picked up on mounting plates riveted to the boiler and firebox castings. It was this arrangement which gave the Foden its great strength with low weight, and Foden sued Wallis & Steevens.

What followed can seldom, if ever, have been repeated, for numerous learned law-lords found themselves discussing in the greatest detail the niceties of engineering practice, stresses in riveted structures, the various means of attaching beam members to cylindrical members, etc, and all at enormous length. What it came down to was whether a saddle which passed under the boiler on the Wallis was a part of the supporting structure for the boiler, or was there mainly for the purposes of supporting the forecarriage. Fodens' lawyers argued that that saddle was part of the undercarriage, and the boiler was primarily attached to the frame rails by the riveted brackets at the side. Seeing a technical way out, the Wallis lawyers argued that the saddle under the boiler was part and parcel of the means of attaching the frame to the boiler. At the first hearing Edwin Foden lost his case, and the court found for Wallis & Steevens, with costs against Fodens. But Edwin wasn't satisfied and he appealed; mistakenly as it turned out.

There had been some *prima facie* evidence that from a pure engineering point of view the Wallis method did in fact use the attachment of the side rails in a very similar manner to Fodens, despite the presence of the saddle at the front, and at a lower court that point might have been forced home. But as the appeal went to higher courts, and finally to the House of Lords itself, there was less and less of an engineering view, and more and more of a purely legal view applied to the case. The transcript of the House of Lords appeal hearing ran to 155 pages of detailed arguments about all manner of engineering design points, and there were bundles of drawings and photographs of supporting evidence; but in the end it was a legal and not an engineering argument that lost the case for Edwin Foden. This was in 1908, and despite the fact that Fodens Ltd and Edwin Foden — which was how the appellants had been named — lost their case, the whole business had resulted in a considerable amount of publicity, and the various stages of appeal had been widely reported in the newspapers. Being a considerably larger firm, Fodens Ltd was able to capitalise on that publicity, whereas Wallis & Steevens Ltd could not. Consequently, although Wallis & Steevens were the declared victors of the contest, it was Edwin Foden who won in the long run, and that is typical of the man. Very few people could ever claim to have 'put one over' on Edwin Foden.

There was a major factor in the design of wagons that was not included in the HMCO regulations, but which nevertheless had a tremendous influence on the characteristics of the wagons then being built. That extra factor was roads, or to be more precise, road surfaces. The vast majority of surfaces all over the country were what was called macadam-surfaced, not to be confused with tar macadam. Macadam meant a mixture of small stones laid over larger stone foundation layers, all bound together with a mixture of clay and water. It was in fact very similar to much of the present rural road network in Scandinavia, and can still be found in the remoter parts of Wales and Scotland. This type of surface was quite durable for light traffic, but under the onslaught of the eight-ton axles of the steam wagons it broke up rather rapidly. The mechanism of the action of heavy wheels on these macadam surfaces was researched thoroughly by William Norris in 1906 in his book *Modern Steam Road Wagons,* and the theory of Norris' work was that, where a wheel was making a rut in the surface, that wheel was in effect constantly having to climb a ramp out of the rut, so absorbing a great deal of power. For example, a 2ft 6in wheel making only a 1-inch deep rut had an effective 18 per cent gradient constantly facing it, and that required a total of something like 10 hp in a wagon weighing about ten tons. So all that power was absorbed by the road surface, in addition to the power required to move the machine along.

Because of this a great deal of experimenting was done with various types of wheels, not only by Fodens but by other makers too. At Elworth the steel-tyred traction-engine type of wheel remained favourite for quite some time, because it appeared to give a better performance over the ubiquitous macadam surface. Very few local authorities were prepared to spend much money on road surfacing, except some of the more far-sighted industrial and merchant cities, particularly in the north-west of Britain, where the use of granite setts to provide a hard surface was common. Wooden wheels, notably by the firm of Bauly's were in favour for a time, because it was felt that they had more resilience, even when fitted with iron tyres, than the traction-engine type.

Slowly road surfaces improved, helped only marginally by the grand-sounding but largely ineffective Road Board Scheme introduced by Lloyd George in his 1908 Budget, whereby greater taxes were levied on road vehicles for the purpose of building and repairing roads. Chairman of the new Road Board was Sir George Gibb whose father had worked with Thomas Telford on many great road schemes a generation earlier, so he ought to have been a good man for the job: and so he would have been had the Treasury let him, but political purseholders being what they are — and were then — the necessary funds were seldom forthcoming at the right time, because they had been 'borrowed' for some other purpose. Nevertheless, road surfaces steadily improved, more by good will than by good management, and by the time Edwin Foden died in 1911 there were sufficient good roads to enable the Foden engineers to think about making faster wagons to run on rubber tyres. Certainly on the old macadam loose-surfaced roads the twin compound would not produce enough power to carry a fully loaded wagon at much more than 6 or 7 mph, and on the granite setts or blocks the shattering ride characteristic was an adequate deterrent to high speeds! But once the use of asphalt or tar macadam became common, as it increasingly did from about 1910 onwards, the way was open for further wagon development. From that time most Foden wagons were produced on cast-steel wheels with characteristic Y-shaped spokes and were shod with solid rubber tyres. This enabled the wagons to proceed at 12 mph if unladen or partly laden, which was a handy legal step. In fact it was found that a great number of them travelled at such speeds or more, even when fully loaded and drawing a trailer, and prosecutions were widespread.

As a result of all this progress on the roads, a 'second generation' five-ton model was produced from about 1911. The road wheels were smaller in diameter than before, the footplate was redesigned with a better driving position, the engine low-pressure cylinder was enlarged slightly, boiler pressure raised by about 5 psi to 210 psi, there were improved brakes and springs, and the body was enlarged again by a few inches. It was with this model that Fodens Ltd faced World War 1. Many machines actually went to war in France, either commandeered from civilian owners, or bought directly from the factory by the War Office. Pickfords Ltd bought 50 of this type during the war to add to their fleet, and the reputation of the wagons there was first class, even though Pickfords were already widespread users of motor vans and lorries.

So successful was this developed five-tonner that it carried Fodens Ltd to their most profitable period of the steam age. The 'five-ton' label was something of a misnomer, because these tough little wagons were quite easily capable of carrying seven tons and, as that still made them just about legal as far as weight was concerned, most of them ran with that payload, or even more, for the majority of their lives. After the model was introduced, many owners of earlier types brought their obsolete models back to Elworth to have them rebuilt to the up-to-date specification, or at least to have those parts of the specification embodied that would enable them to take full advantage of the raised speed limits. This rebuilding activity, in addition to the building of new wagons and experimenting with ideas for further types, kept the works at Elworth extremely busy when many other factories in the heavy vehicle industry were on short time or, even worse, closing down.

The main reason for this surge in activity at a time when many others were heading in the opposite direction, was once again a matter of

*A symphony in steam*

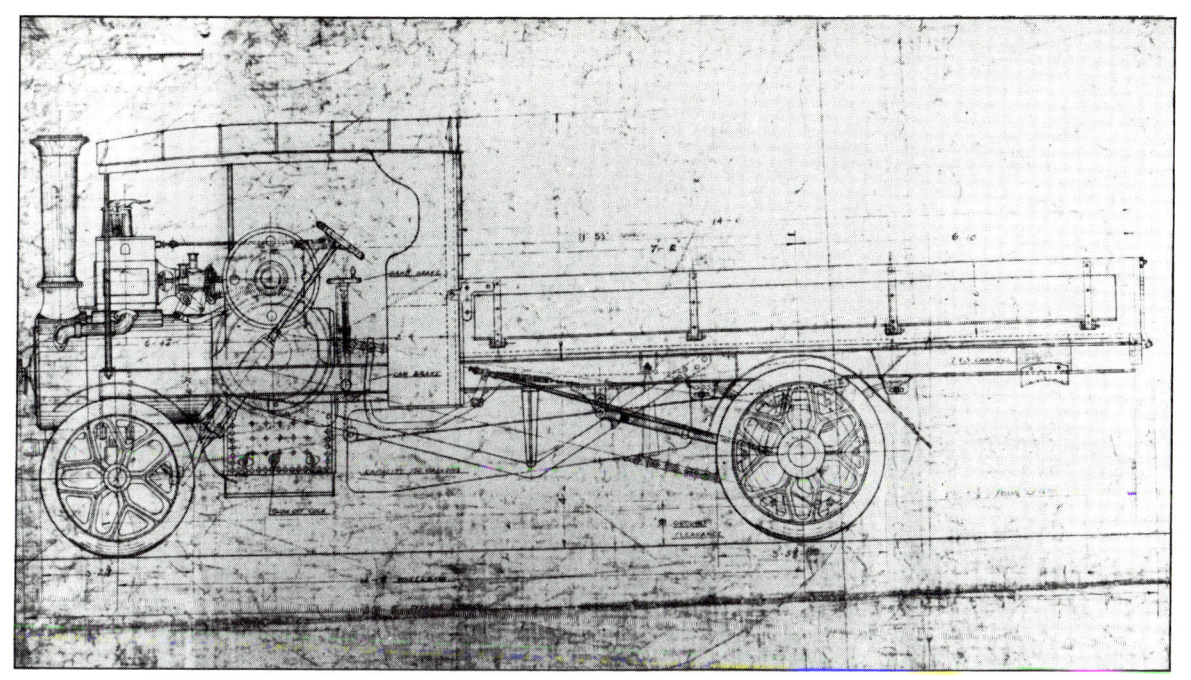

*The C-type six-tonner was a refined version of the five-tonner with bigger loadspace, higher speed and Ackermann steering. It was undoubtedly the best overtype ever produced.*

legislation rather than any technical or operational factor. From 1908 onwards, when Lloyd George as Chancellor of the Exchequer proposed a tax of 3d on a gallon of petrol, successive Chancellors had sought to make their names as fund-raisers, and each in turn added a few pence per gallon. This went on right through the war years until, by 1920, the price of motor spirit was something like 4s 3d a gallon. This was an exorbitant price for those days, especially in view of the fuel consumption of a heavy petrol lorry, which would be between 3 and 5 mpg, depending on the type. Coal on the other hand was cheap, and it was plentiful. The way of the petrol-powered lorry was not smoothed by alarmist notions widely publicised at the time, to the effect that so much petrol had been consumed by vehicles during the war years that the entire world supply would only last a further three years at most. This still sounds a familiar kind of story today, but the threat seemed real enough in 1919-20 and so the steam wagon was more popular than ever before, despite its relatively high price.

A Foden five-tonner at that time cost £900 give or take a few pounds, depending on the type of body and fittings that you ordered, while a new, heavy five-ton Leyland, for example, could be had for about £560. On the other hand, there were increasing numbers of reconditioned ex-army lorries, from about £300 upwards, some rebuilt by their original makers and consequently very good value, others less skilfully attended to by back-street 'pirates', and of rather doubtful value. Despite this level of competition from the heavy petrol lorry the Foden steam wagon retained its enormous popularity. In any case, there had grown up around the steamers a strong atmosphere of loyalty, almost a mythical status, and the experienced steam hand was revered among lesser transport men.

At Fodens Ltd things seldom stood still and, despite the success of the ubiquitous five-ton model, the brothers E.R and Willy Foden had already completed a great deal of experimental work aimed at a further improvement in carrying-capacity and efficiency. The new model was first mooted at the annual general meeting in 1922, when 'our new six-tonner' was mentioned as a possible saviour of trading fortunes in an economic climate that was quickly deteriorating. It was not until the following summer that the six-tonner appeared as an official new model, although versions in various stages of development had been seen on the road before then and had even been loaned to operators for field trials. The newcomer was named the 'C-type'. Just why is not clear, as no letter designations had been used previously, but there are distinct references to C-types in all sorts of records and documents from 1923 onwards, whereas all the early references to the machine in its development stages simply referred to it as the six-tonner. It was probably a decision made on the spur of the moment by the sales side of the company, as number or letter designations were becoming a popular method of identifying vehicles in the early '20s, as the multiplicity of types and models grew. Some makers like Leyland and Dennis had such designations from a very early stage, partly because their model ranges were more diverse than most, but Fodens had never before needed to do so.

Whatever the reasons for its name, the C-type became an almost overnight success. Because of the economic climate, which was taking Britain towards the biggest depression in its history, sales had been increasingly hard to come by. As one annual report at Elworth put it, 'The trouble with the Foden Wagon, if it can be so called, is that it does not wear out quickly enough. Consequently when times are hard the owners keep on running their old Fodens instead of buying new ones.' This made an original excuse to offer the shareholders as a reason for falling sales, but it was not the only reason. The tax on petrol had been drastically reduced as the Treasury discovered that they had sucked their 'golden goose' dry, and by 1923 the price was down to about 1s 4d a gallon again, roughly the same as it had been before the war. This made the petrol lorry a much more attractive proposition, and that included the rebuilt army types. Against these price levels the expensive steamer was hard-pressed to compete.

The C-type, however, gained a brief respite for Fodens, despite all the opposing factors. It was fast, it was economical, it had three speeds, a better cab, a larger loadspace, and its price was little more than that of the five-tonner at £940. So well did the haulage industry accept the newcomer, and so many did they buy, that the wage cuts and redundancies that had been threatened the year before were unnecessary, and an increased level of trading and profit made things look good, or at least better than they had done since 1920. The C-type Foden is generally supposed to have been the zenith of design and efficiency in overtype steam road-wagons, so a closer look at some of its technical features is warranted. It still had the locomotive-type boiler, although other steam manufacturers including Leyland, Yorkshire and Sentinel had all forsaken it. But it did have its advantages, among them the ability for the engine to be mounted directly on top of the boiler, so keeping steam pipes short, and enabling the cylinders to drain directly into the boiler. That loco boiler had 90 sq ft of heating area, gained from 63 fire tubes. These tubes were cold-drawn from a special chrome steel, and were 1 3/8 inches in diameter. The boiler-barrel was 26 inches in diameter and a superheater was added in the forward part of the boiler in the smoke box. The fire grate had an area of 3¼ sq ft, and the working steam-pressure for normal use was 220-240 psi, although each and every boiler was tested up to at least 380 psi before leaving the works.

The engine was basically the same double-crank compound design that had been developed for the traction engines in 1880, more than 40 years previously. But it had undergone a great deal of refinement in the meantime and now, with a 4¼-in diameter high-pressure cylinder, and a 7-in low-pressure cylinder, the engine developed 25 hp at about 450 rpm. Foden patent double-D slide valves operated by Stephenson-Howe link motion, functioned in outside steam chests and the cylinders were fully steam-jacketed. Cast-steel gears were machine-cut and provided a much quieter means of transmission than the older two-speed types. As before, a Renolds chain drive to the differential on the rear axle was used and, with standard sprockets, the three speeds gave theoretical road figures of 6 mph, 12 mph and 16 mph. There were, however, optional sprocket sizes to be had at the Foden service stores, and many operators chose these to achieve speeds of up to 25 mph or more,

*A symphony in steam*

although this was way over the legal limit.

Such unprecedented speeds on a heavy steam wagon were relatively safe, because the C-type had very good brakes by the standards of those days. There was a cam shoe brake operating on the rear wheels from a pedal which could be locked down, and in addition there was a band-type brake operated by a hand-screw. Of course, as in all steam wagons, the 'engine reverse' system of braking was also available. Earlier Fodens had an open-sided cab, with the driver perched on a wooden seat above the left-hand frame member with a foot-rest below, and this later developed into a sort of sidecar structure built on to the side of the wagon. This can be clearly seen in some of the photographs in this book. Fodens introduced a cab with fully enclosed sides in the C-type, with a door for access, and soon after its introduction a windscreen was made available. This was partly the consequence of higher speed, because at 8 mph —

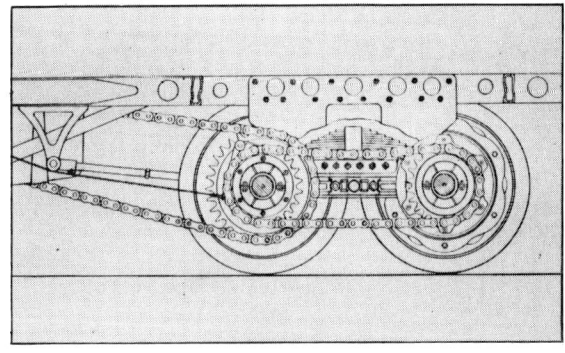

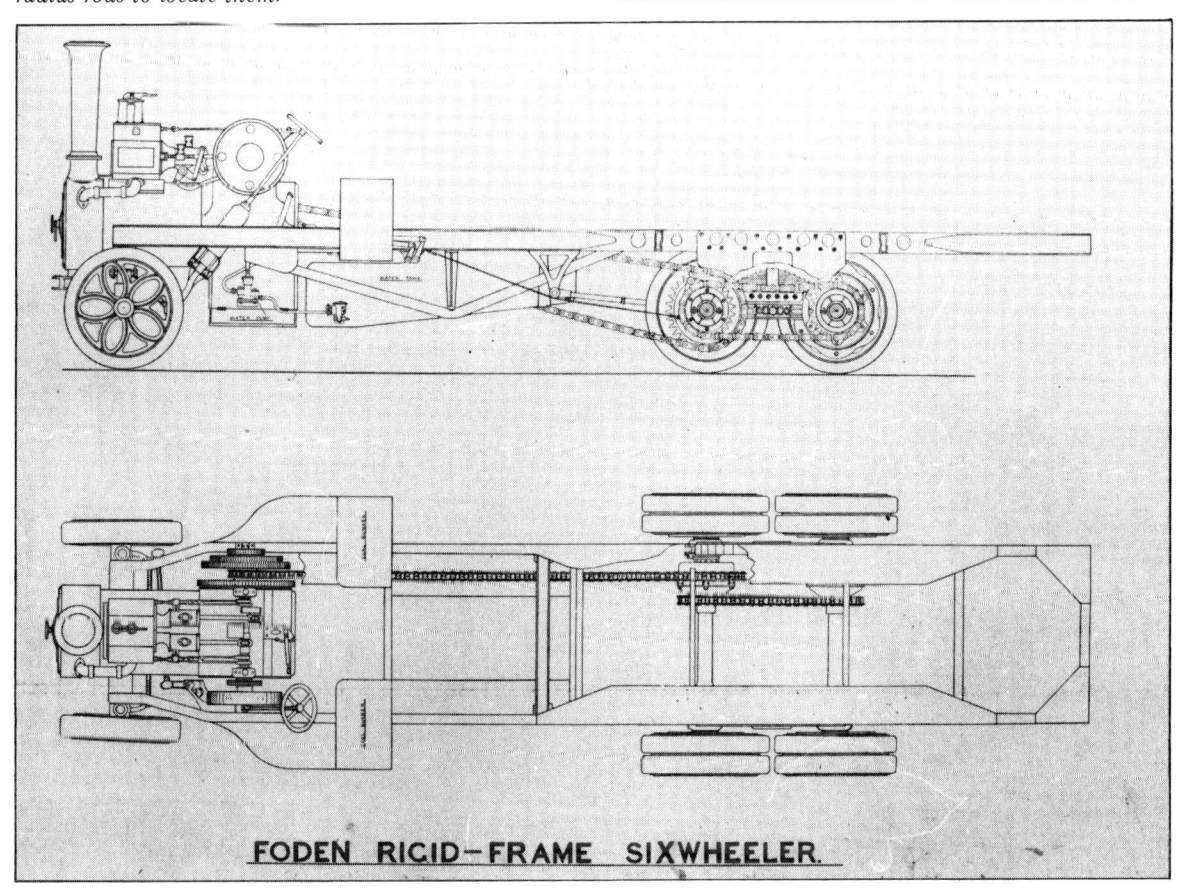

*The addition of a tandem bogie turned the C-type six-tonner into a genuine 10-12-tonner. The bogie had a surprising degree of articulation. Inverted trunnion-mounted springs carried the axles, with short adjustable radius rods to locate them.*

FODEN RIGID-FRAME SIXWHEELER.

66     *The Foden Story*

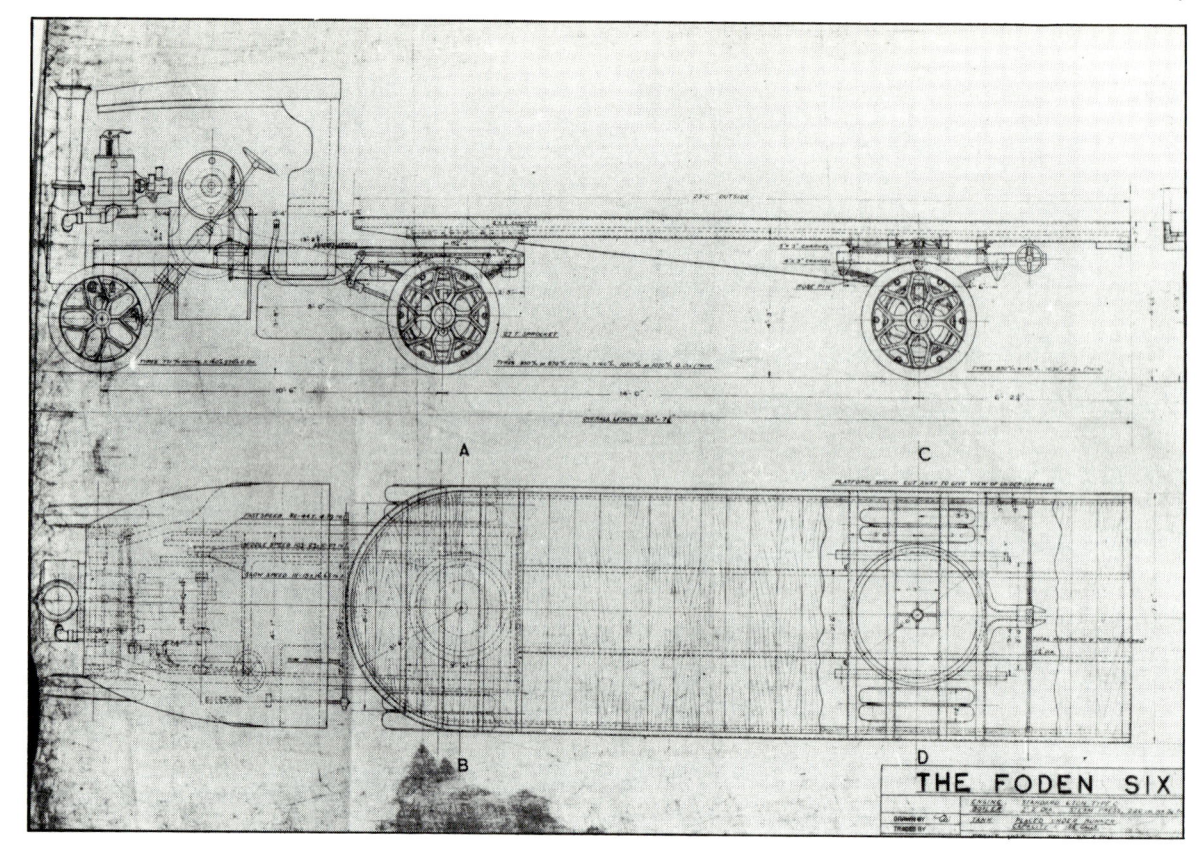

*A 'flexible six-wheeler', or artic, was developed from the D-type tractor. Novel features included a spherical coupling plate and a steering trailer axle.*

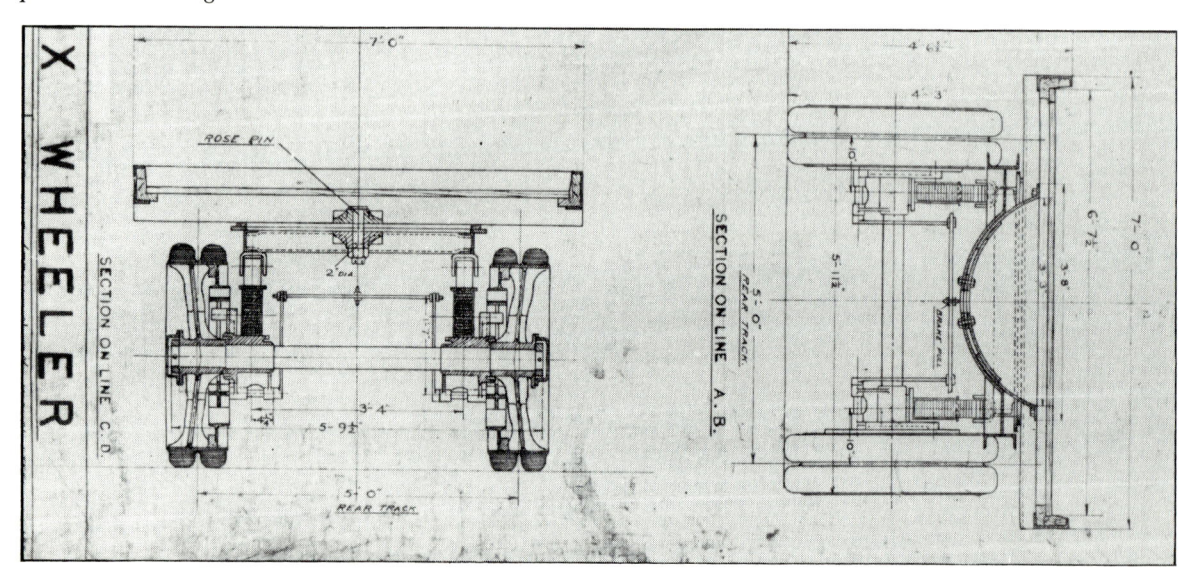

even in heavy rain — very little discomfort was felt by the crew back under the canopy, whereas at 20 mph-plus, a great deal of rainwater could penetrate the cab in bad weather. Even so, many of the old hands despised the notion of a windscreen for many years after it became available.

Speed capability also meant that something had to be done about steering gear. A rotating forecarriage carrying the front axle and spring, all controlled in a rather haphazard manner by a bobbin and chains, was adequate enough at 8 mph, even if it was basically the same system that had been used on the first traction engines over 40 years earlier, but with higher speeds greater accuracy was necessary and Ackermann-type steering gear, where the wheels turned on individual kingpins on the fixed front axle, provided the necessary degree of control.

The standard wheelbase of a C-type was 14 ft 9 ins, within an overall length of 23 ft 6 ins, and the usual body dimensions were 14 ft 6 ins long and 7 ft wide. The road wheels were almost the same size front and rear, except that the rears were twins. The latter were 3 ft 5½ ins in diameter over the tyres, and the fronts were 3 ft 2¼ ins. There were several variations of the C-type, including a variety of wheelbases, some with hydraulic brakes, and a few with fully enclosed cabs. In later years a few were built with pneumatic tyres, and large numbers were converted to these after the Salter Report imposed high taxes on solid-wheeled vehicles, but we shall come to that particular phase in history in due course.

The higher performance and capacity of the C-type enabled steam-wagon operators to keep pace with the heavy petrol-engined lorry which by the mid-'20s was making great strides in economy, reliability and speed. The C-type's engine had so much pulling power that it was easily able to haul higher weights than the law allowed a four-wheeled wagon and, although a great deal of overloading went on among hauliers when they thought the Law was not watching them, this was not an entirely satisfactory solution.

The next step, just one year after the C-type itself appeared, was the 12-tonner. In effect this was a standard C-type front end, with a tandem bogie at the rear. The drive to the bogie was by Renolds chain as before, and a second chain linked the two axles of the bogie. The suspension was ingenious, with inverted leaf springs mounted on trunnions at their centres, the axles riding on hardened, formed seats which fitted the shaped ends of the springs. Radius rods gave additional location, and despite the rather cumbersome appearance, each axle could articulate independently of the other as much as ten inches above or below its normal position. The geometry was so well worked out that this displacement did not disturb the running of the chains. To achieve the required degree of safety and security, big cam-operated brakes were fitted to both axles of the bogie, pedal operated from the cab. Although Fodens patented this bogie in 1925, it did not appear to stop others from making a very fair copy of it.

The engine of the 12-tonner was identical to that of the six-tonner, but the boiler ran at 230-250 psi and the power output was increased accordingly to 28 bhp. The gearing was slightly lower too, and operators were not quite so fond of installing 'fast' sprocket sets on 12-tonners as they were on the smaller wagons.

On its factory trials, the 12-tonner was taken to the hilly country to the east of the Cheshire flatlands where it had been built, and was driven up and down every available hill road until the engineers were satisfied that it would do all that they claimed for it. In particular they did a whole series of restart tests on hills such as Mam Tor and Mow Cop, neither of which had asphalt surfaces at that time. In fact, Mam Tor was still loose surfaced after World War 2. Mow Cop was the scene of a great number of tests over the years, notably when both Fodens and ERF, their eventual neighbours, were developing air-brake systems for diesel lorries in the '50s. It is still a dangerous hill, but how much more so it must have been in 1925! There was plenty of power available for fully loaded restarts, even on Mow Cop's 1 in 4 gradient. Traction was something of a problem, however, and subsequently special chain-secured grips were developed to fit on the smooth exteriors of the solid tyres.

These 12-tonners were very popular with the stone-quarrying and road-building industries, where bulk meant profitability, and they could be seen chuffing around all the major quarrying districts of Britain such as the Mendips, Peak District, Cotswolds, West Cumberland and Wales, for many years after their introduction.

Another development of the C-type was the 'flexible six-wheeler', which first appeared in 1924. In modern times we would call this an articulated truck, but the term was not used at the time. The tractor portion of the vehicle was virtually a standard C-type chassis, but it was fitted with a spherical type coupling to which was attached a long semi-trailer. The trailer was not easily detachable in the manner of the modern artic, but the idea was to provide a large loadspace for cargo, mainly for the distribution of manufactured goods and foodstuffs — an increasingly important function for the heavy wagon by the mid-1920s. Clearly the Foden engineers were concerned about manoeuvrability because, even though the trailer was only 23 ft long, it was provided with a steering trailer axle to assist it round the tighter corners. In practice this was seldom used, in fact the 'flexible' itself never became particularly popular, and only a handful were made.

*The D-type tractor, built for trailer haulage, was virtually a short-wheelbase C-type. Later many C-type wagons were shortened to tractors, which explains why so many seem to have survived.*

Yet another variation on the C-type design was a short tractor called the D-type. Instead of a load-carrying body, the D-type had a ballast box over the rear wheels and, although most of them had larger rear wheels than the wagons, this was not always the case. D-type tractors were popular in freight yards, dock areas, and places like steel works, where they were employed to haul trailers around, mainly for short distances. Having a shorter wheelbase than the wagons they were surprisingly manoeuvrable. A handful of D-types were turned out as showman's tractors, for customers who had in years gone by purchased showman's version of the big traction engines. The showman's D-type had a decorated and extended canopy, and the usual DC generator mounted ahead of the smoke stack, where it was belt-driven from the flywheel once in position on a fairground. It could therefore provide ample supplies of electricity for lighting and ride-power, at a very economical cost. A number of Ds were fitted with the more powerful engine from the 12-tonner, which made them capable of pulling several loaded trailers at once. There is on record a D-type which habitually hauled over 40 tons gross in East Anglia, working for a timber merchant. The tractor worked with two big timber drags right out of the woodland. It would haul or winch each laden drag out of the wood individually, then couple them up and tow the whole lot as much as 30 or 40 miles to the sawmills. This of course required a great deal of driving skill, a topic dealt with in the following chapter. Innumerable legendary tales of this calibre exist about the overtype wagons, in particular the C and D types and the 12-tonners, and it is little wonder that when the overtype, with all its disadvantages such as wasted space and tricky boiler handling in hilly areas, finally gave way to the undertype which was theoretically superior, there were many old hands at the steam game who reckoned that steam haulage would never be the same again. That curiously English habit of looking back and yearning for things obsolete undoubtedly had something to do with these sentiments but, on the other hand, with the benefit of hindsight there can be no question that with those magnificent overtypes of the mid-'20s, Elworth's symphony in steam reached its crescendo.

# Chapter 8

# Men of steam

Driving a steam wagon was a relatively straightforward business. Driving it well demanded a great deal of skill, expertise, patience and sheer cunning. Any steam man will tell you that a wagon was not simply a lump of machinery, but a living being, with a mind of its own, and probably with as many sly tricks as an old fox. The idea of good driving was not to let the machine outfox you, but to keep at least one step ahead for as much of the journey as possible. The mere act of starting a steam wagon from cold called for no little patience and perseverance, certainly a great deal more than most modern drivers would possess. The following instructions are taken, in shortened form, from various Foden handbooks for wagon-drivers.

'The boiler should be thoroughly washed out by removing the boiler barrel manhole cover and firebox mudhole doors, and completely flushed through with a hose, at as high a pressure as possible. The top of the firebox can be cleaned with a stiff brush at this time. When all is clean, replace the manhole covers and make sure that they are put back squarely using the proper oval steam joint rings covered with a light smear of graphite. Next the feed water tank should be flushed through in the same way. Attention should be paid at this time to the strainers at the pump and injector suction pipes, and to the feed water check valves on the side of the boiler. These must be seating well, and free from dirt. The water gauges must then be cleared to ensure that the cocks are not blocked and the glass clear. The boiler may now be filled, with rainwater if at all possible, using the filling plug. Fill until the water level shows about three quarters of the way up the glass. Do not overfill. This may cause entry of water to the cylinders and consequent damage. After the boiler is filled, the feed water tank may also be filled, and while that is being done the preparation for the fire may be done.

'Soot and dirt should first be brushed from the boiler tubes, smokebox and firebars. Lay the fire by placing a thin layer of oily waste or woodshavings on the bars, then cover that with a layer of dry wood, with a thin layer of coal on top of that. Break the coal, preferably Welsh steam coal, into pieces about three inches across. Make sure that the boiler is properly filled and the filling plug tightly replaced, before lighting the fire. Raise the ashpan lid to admit ample air at this stage. While the engine is steaming up, it is good practice to go right round the engine and see that all lubricators are correctly charged with the type of oils shown in the lubrication chart. When steam pressure is part way up the red mark on the gauge, open the cylinder drains fully, and just "crack" the regulator open a fraction, to warm the cylinders through. Blow the gauge glass down to ensure a correct water level reading. (This entailed opening and closing the steam and water cocks in turn, then in reverse order.) If a superheater is fitted make sure the stop valve is open, then blow the superheater down thoroughly before starting off, until dry steam issues from the drain.'

All this performance was just to get the machine ready to move. An experienced steam man with an engine in good condition could do it in half an hour or so, but a novice could take three times as long. It was difficult to do the preparations quickly on an old engine, too, especially if the tubes leaked where they showed through the tube plates at the ends of the boiler. Tubes could be fitted with little tapered ferrules in their ends, and if a driver saw the tell-tale 'tears in the eye', in other words water leaking from the tubes, the ferrules could be tapped in a little more to tighten the tubes in the plate. If left to 'weep', rapid damage would occur to the joint between tubes and tube plate.

Driving the machine entailed the use of three main controls, the 'reversing lever', which moved in a quadrant at the left near the flywheel, the 'steam regulator' over the centre of the boiler, and the 'three-way cock', again at the left below the reversing lever. To move forwards the reversing lever was pushed right forwards in its quadrant, the regulator was eased open, and the wagon would move off. If it did not it usually meant that the pistons were in the 'over-centre' position, and the three-way cock then had to be pulled fully backwards to change the relative position of valves and pistons. This device was originally developed for starting heavy traction-engines in the 1880s. As soon as the engine turned, the cock was returned to its normal place. On the level roads with a reasonable load, the wagon would quite happily move off in high gear, bearing in mind that, unlike an internal-combustion engine, a steam engine develops its maximum torque at minimum speed. But for hill-climbing, manoeuvring, and very heavy load-pulling on the level, low gear had to be used. It was not possible to change gear on the move, but once the wagon had stopped the gears could be slipped easily into or out of mesh by the slide control to the right of the regulator. This operation had the advantage that the driver could see exactly what he was doing, as the gears loomed large and greasy, right in front of his nose!

Keeping the correct level of fire and pressure of steam for all eventualities was where the real skill of steam-wagon driving came in. To begin with, it was essential that you knew what the road was like on the route you were about to take. If there was a long hill, for example, you had to start preparing for it long before it came into sight. Fire would have to be built up so that a full head of steam could be maintained, even though the engine was taking as much steam as could be fed to it in order to climb the hill. But once near the top, you had to prepare for the descent where very little steam would be needed, and so the fire would need to be slowed down in good time so that no excess steam was generated. It was an admission of defeat to a real wagon man if he allowed so much steam to build up that the excess had to be blown off.

Another skilful job was keeping the boiler water-level correct. There were two ways of doing this. One was to use the pump, situated at the right of the firebox, and that was normally used with the wagon stationary or working under light load. At all other times the injector was used, and this was at the left of the firebox. The injector was, in effect, a steam venturi valve and, if you opened the steam valve on it, followed immediately by the water valve, the steam would blow through the injector, taking water with it into the boiler. Because of the high pressures and temperatures in the boiler, great damage could be done if the water level dropped too low, and once below a certain level there was no way you could get it back without shutting down the boiler, releasing all the steam, letting it cool off and starting again from cold. If a careless driver allowed his water to get too low, there was a device called a fusible plug in the firebox crown, normally covered with water, of course. If it was ever uncovered it would, as it was made of lead, fuse, and release the boiler pressure to avoid damage. The man who ever let his fusible plug go was in disgrace indeed, as the wagon would then have to be allowed to cool thoroughly, and a fitter would have to be called out to install a new fusible plug.

Operating in hilly country was tricky, quite apart from keeping the required level of steam pressure and fire. Foden wagons were particularly sensitive to gradients because of their long locomotive boilers, since if the water level was right at the end of the boiler then it would be wrong at the other end all the time the machine was on a slope. (The main advantage claimed by Leyland, Thornycroft, Yorkshire and others for their vehicles, was that vertical or transverse boilers were not so sensitive to slopes and were therefore easier to handle.) Keeping the fire properly stoked was an art in itself. The aim was to keep the fire as even as possible right across the grate, with perhaps a thicker layer of coal around the outside edges. It was important not to let excess air spaces burn through the fuel layers, as this would produce relatively cold air-currents impinging on the area immediately above, giving rise to distortion. Fuel was usually added in rotation, such as rear-left, front-left, rear-right, front-right. The firebox door had to be kept shut when pulling hard, again to avoid cold air getting in among the hot metal, so experienced wagon men always stopped to make sure their fire was in good order before starting a long hill. If any extra fuel had to be added on the way, it was popped in quickly with a minimal opening of the firebox door. Each engine man developed his own style of doing things, and long discussions would be held at the end of the day as to which was the right and which the wrong way to tackle a particular stretch of road. Even so, they would probably all be right, as every wagon had its own idiosyncracies. It was considered bad form to add fire coming up to town; the proper way was to build the fire up well before getting there and then drive through with a minimum of steam and no smoke. In rural areas where there were thatched roofs, haystacks, and the like, the damper had to be kept shut to avoid sparks or scraps of hot cinder straying from the chimney and landing on inflammable material. Public opinion always remained sensitive to the effects of steamers.

Fodens always recommended that a wagon have one day off a week for maintenance purposes. When a wagon was left for the night, they recommended that it be sheeted over, the reversing lever set in the 'mid-gear' position, and the fire left to die slowly. The mid-gear position was the notch next but one nearest to the driver on the quadrant, and it set the valves so that there was steam on both sides of the pistons at once, forming a kind of brake effect. On the road this was indeed an effective engine brake while descending hills.

The normal maintenance check list was relatively simple, but left little to chance. It gave the following instructions:

*An early five-ton chassis, with William Foden in command, pauses outside the new office block for a photograph.*

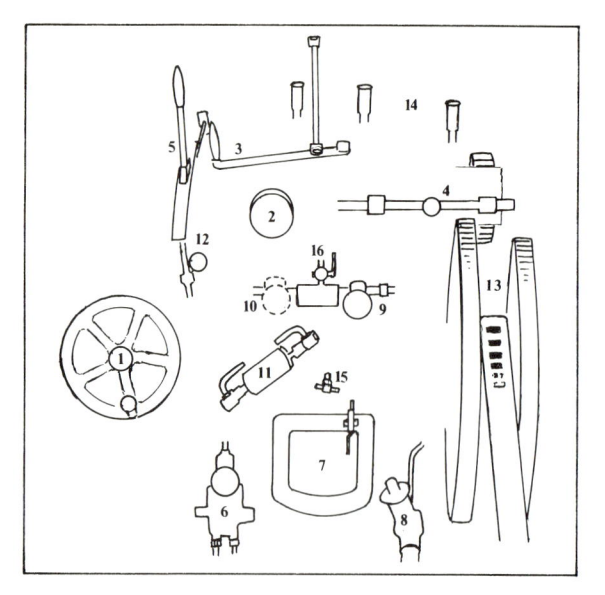

*This view of a five-tonner with cowling removed shows the main controls, identified in the sketch. 1—steering wheel; 2—pressure gauge; 3—regulator; 4—change speed; 5—reversing lever; 6—injector; 7—firebox door; 8—water lifter; 9—lifter steam-valve; 10—injector steam-valve (missing in this case); 11—level glass; 12—3-way valve; 13—gears and chain; 14—lubricators; 15—filler plug; 16—cock for gauge.*

1. Check all glands and packings, tightening or repacking as required.
2. Check that all steam joints are tight.
3. Flush out boiler at 250 mile intervals, or more frequently if dirty water was used.
4. Clean the water feed tank every three months.
5. Boiler maintenance as follows:
    a   Do not blow down boiler with fire burning
    b   Caulk any boiler leaks with a proper caulking tool
    c   Expand any leaky tubes with the ferrules provided
    d   Any burst tubes should be plugged until they can be replaced
    e   Remove boiler lagging annually to clean and repaint boiler plating
    f   Do not refill boiler while it is still hot
6. Clean all soot from superheater coils if fitted.
7. Check safety valves periodically, noting blow-off pressure.
8. Lift safety valve handle occasionally to prevent sticking in place.
9. Check condition of all brasses (bearings) making sure they are neither loose nor tight.
10. Clean chimney interior, check that exhaust cone is in true centre.
11. Steering chains and buckles must be correctly adjusted, and tight.
12. Rear axle radius rods must be correctly adjusted to give correct chain tension and keep axle parallel.
13. Drive chain should be removed occasionally and thoroughly cleaned with kerosene, relubricate it by immersing in hot tallow or a graphite base chain lubricant.
14. Drain the boiler and water tank if laying the wagon up for any length of time, ease off all bearings and brasses, cover the engine with a sheet, protect bright metal parts with a preserving oil.

There is no doubt that steam wagons were robust machines and very durable, provided they were properly looked after. But they could very quickly make a monkey out of the impatient, unskilled, or simply stupid operator who declined to heed the little signs and signals that every wagon gave its crew.

Usually there was a crew of two on a wagon, a driver and a fireman. It was not a question of who was the boss, because it was very much a team effort, but in practice the driver was usually the senior man. The overtype Foden had the inestimable advantage that it could be operated quite easily by one man, providing that he knew his road, the loads were not excessive, and he was familiar with the tricks of his particular wagon. On the three- and five-tonners he sat on a little wooden seat above the left hand frame member, with one leg outside, the steering wheel low down in front of him, and the reversing lever and regulator within easy reach. It was only a short reach to the change-speed lever, and the firebox door was to his immediate right, with the pump and injector ranged on either side of it. Later six- and 12-tonners had a wider cab with a higher floor, but they could still be operated by one man. However, the undertypes which came still later had to have a two man crew, for the simple reason that the driving compartment was on the right, and the firebox and so on at the other side with the boiler in between, and there was no way any man could attend to both sides at once. For many, this was a retrograde step, despite other advantages possessed by the undertypes.

Many were the stories told by the wagon men of their adventures on the road, for such was road haulage in its formative days that nearly every journey was an adventure in itself. They were pioneering days in every sense of the word. Dire tales were told of what would happen if maintenance were neglected, and frequently the worst actually happened. Take, for example, the plight of a Sussex owner of a Foden three-tonner, purchased second-hand to carry vegetables and farm produce to market, just before World War 1. Being a man of the soil, he was not particularly mechanically inclined and had taken the vendor at his word when he said that, 'these wagons are so good they nearly look after themselves'. The new owner discovered that this wasn't quite true one summer afternoon when, chuffing gently down to a farm below Arundel to pick up a fresh load of cabbage, he suddenly found himself travelling along the main Southern Railway line in the direction of Portsmouth, with a train coming the other way! Tough though the Fodens were, this would have been one battle almost surely lost had it ever been enjoined. Fortunately the protagonists both stopped in time. What had happened to the Foden was that the steering chains had been allowed to get very loose and sloppy. The front wheels and axle therefore rocked as they traversed the wooden sleepers of the diagonal crossing over the railway, the whipping chains flipped one of their shackle pins right out, allowing the forecarriage to swivel freely. The front wheels simply followed the line of the railway, taking the rest of the wagon with them! The incident must have had some beneficial effect on the owner concerned, for he ran his wagon for many years subsequently with no further disasters, as far as we know.

Some users simply neglected their machines to the point where they were practically inoperable, despite the ruggedness of the design. Such a case was reported by one of Fodens' own drivers, Ken Judkins. He had delivered a new wagon to the dealers in Inverness, and was asked to take back an old wagon which had been lying dormant for about 18 months after approximately ten years of hard use. Outwardly it was very dirty, but it did not look too bad after a clean down. Routine washing-out of the boiler and water tank; fetching some fresh water in milk churns on the local coalman's petrol truck; making some temporary joints with asbestos rope and red-lead putty; and buying a small shovel for the fire at the village ironmonger's shop — all these were necessary before any steaming could be contemplated. The first warm-up to about 50 psi in the boiler showed no great shortcomings. After everything had been oiled that could be oiled, and the flywheel had been turned over by hand a few times to make sure everything was free, the regulator admitted real steam and all worked surprisingly well. After a trial circuit of the village, when all again seemed well, and a quick sojourn at the pub to take in extra courage for the journey, Ken took the old three-tonner on the road for the first time in a year and a half, and in fact made about 35 miles down to Stirling that afternoon without trouble.

At the next morning's inspection some tubes were seen to be weeping, and by the time 150 lbs of steam was up, about ten of them had 'tears in their eyes'. A little work knocking in the ferrules dried them all up, except a couple of persistent ones, and the old wagon hit the road once more, heading south for Carlisle. All went well down through Aidrie and Lanark, but then steam pressure began to fall once again, and yet more tubes were leaking due to the vibration that was slowly but surely loosening all the dirt and scale which was holding the inside of the boiler together. More ferrule tapping restored some of the tubes, but pressure would not go above about 170, and it was supposed to run at 200 psi. Fortunately the coal was very good, so it was easy for a skilled man like Judkins to keep tight control of the fire under the delicate boiler, and the three-tonner set off again. Ken hoped to reach digs he knew at Ecclefechan before nightfall, but the going was so slow that he only reached the tiny village of Roberton, north of Abington.

After a night's rest, and careful attention to the tubes yet again, the ancient wagon set off hesitantly, making only about half speed on low pressure. Then, way out in the wilds between Elvanfoot and Beattock, there was a tremendous clattering and a bang. The rods to the eccentric on the low-pressure side had broken, and there were pieces of loose and bent metal all over the place. Ironically, the boiler was by this time almost completely leak-free and in good operational order! This seemed to be the end of the trip, and so it would have been for most wagon drivers: even today that must be one of the loneliest stretches of main road in Britain. But Judkins had heard older steam men talk of the way a Foden compound could be run on one cylinder, and indeed he recalled reading an old agricultural-engine catalogue that had said just that. But how was it done? And how many broken bits could the old girl put up with before she wouldn't work at all?

Over a pipe of tobacco he thought on the problem for a time, then decided it could be done. All the loose bits, eccentric straps and so on — all of which were fortunately on the low-pressure side — were removed, and so was the low-pressure connrod and crosshead. The valve on the low-pressure side was centred in the middle of its stroke and clamped in position by tightening its gland nuts as far as they would go. Finally the low-pressure piston was locked up in a similar way at midstroke. Gingerly the regulator was opened, and wonder of wonders, it all worked, or at least revolved smoothly. A bit rough at low speed when pulling hard, but it worked well enough. By then it was three o'clock and still over 50 miles to go to reach Carlisle, over hilly roads too. A phone call to the Carlisle agents asked them to stay open until the casualty arrived and in fact, when the old three-tonner finally limped in, boiler shaken about by the rough running, tubes spurting water and putting half the fire out, and only 90 psi of steam left, it was well past nine o'clock, but the entire depot staff had stayed behind to see how you could possibly steam a wagon on one cylinder.

Judkins was hoping for a nice train ride back to Elworth after that, but when he phoned the works he was asked to wait until the agents did

*Men of steam*

**Right** *The more powerful steamers were not without their legal perils, as this 1923 summons shows. Cooke was one of Fodens' leading drivers.*

**Below far left** *Steam adventures were not confined to Britain. This D-type tractor was bogged down in Argentina but eventually pulled itself out two days later.*

**Below left** *Foden steamers were well known overseas. These two — a Speed-Six and a Speed-Twelve — appeared at an exhibition in Buenos Aires in 1930, and considerable numbers of steam wagons and tractors worked in Argentina.*

```
In the County of Bedford,
              Petty Sessional Division of Bedford.
   To Richard Cooke _____
   of 15 Eva Street, Elworth     in the County of Chester
      Information has been laid this day by WALTER GEORGE PURSER of Bedford in the County of
   Bedford Superintendent of Police for that you on the  23rd.       day of  November
   One Thousand Nine Hundred and Twenty Three at  Roxton         in the County of
   Bedford did unlawfully drive a Heavy Motor Car with trailer attached
   at a speed exceeding 5 miles an hour contrary to certain regulations
   known as The Heavy Motor Car Order 1904 duly made in pursuance of
   the Motor Car Acts 1896 and 1903

      You are therefore hereby summoned to appear before the Court of Summary Jurisdiction sitting at
   the Shire Hall at Bedford in the County of Bedford on Saturday the  8th.    day of December
   1923, at the hour of Eleven in the forenoon to answer the said information.
      Dated the   27th.     day of   November       1923.
                                                                          L.S.
                                  Justice of the Peace for the County of Bedford.
```

some repairs, and then drive the wagon down. It took the Carlisle men until 8 pm the next day to repair the broken eccentric rod, fit new studs all round, fit four new boiler-tubes and re-ferrule the rest. Tests that evening showed all to be well, so next morning, with a healthy 200 psi on the gauge for the first time since goodness knows when, the old three-tonner set off to the place of its birth with a happier Judkins at the controls. This was one of the wagons where the chain sprockets had been swopped over at some time in its history and, with a 15-tooth gear-shaft sprocket driving a 30-tooth axle-sprocket, the wagon would comfortably do 30 mph. The trip back to Elworth, which must have been all of 150 miles from Carlisle, was completed in a single day, a sharp contrast to the painful progress of the previous two or three days.

Those 'fast' sprockets were frequently the cause of wagons being stopped for speeding and their drivers prosecuted. One such unfortunate occurrence overtook another factory driver, Richard Cooke, who was on demonstration work in the Home Counties at the time. A letter from the works manager at Elworth to the company solicitors, Poole, Alcock & Co of Sandbach, dated November 30 1923, tells the sad tale.

'Dear Sirs,

'We are sending herewith a Summons issued to driver Cooke who is with one of our Demonstration wagons and trailer.

'The verbal report we have from the driver is that the wagon and trailer were running light from St. Neots to Baldock to pick up a load. Being on demonstration work he was anxious to make the best of his way on the unloaded journey. The Police held him up and asked Cooke if he knew he was only allowed to travel at a rate of not more than 5 miles per hour. Cooke replied that he was aware of the fact. The Police told Cooke the wagon was travelling at about 12 miles per hour which Cooke could not deny. If you require Cooke to be in attendance on the 8th prox, at Bedford please let us know in good time as he is still with the wagon.'

Cooke was one of the unlucky ones. Many were the occasions that those old Fodens made astonishingly fast journeys and, even allowing for some slight 'stretch' in the telling of the tales, there can be no doubt that with the right sprockets fitted very creditable speeds could be maintained.

*Mr Edward Heath flags off Jack Hardwick's 6-ton Foden from Horse Guards' Parade at the start of the 'Drive into Europe' in January 1973, while steam envelops the Household trumpeters.*

Even old *Puffing Billy,* the band bus, built before World War 1, and frequently used as a test-bed for new ideas, was at one stage fitted with roller bearings instead of plain brasses, and had a 15-30 sprocket combination. Charlie Williams, the driver at the time, steamed the old bus complete with the band, instruments and music, from Sandbach to Chester in exactly one hour, and that is a distance of 27 miles. Another driver, Freddy Goddard, who also worked with Hillhead Quarries of Buxton, for many years staunch users of Fodens in both steam and diesel forms, claimed that with the six-wheeled 12-tonners he could make it from Buxton to the factory at Elworth in under an hour. That route is well over 20 miles, and what is more it has the notorious 'Cat and Fiddle' hillpass to negotiate on the way. The run down those western Pennine slopes must have been quite exciting!

But adventures with steam wagons are not confined to days gone by. In 1971 a young enthusiast, Mike List-Brain, set off on a round-the-world tour driving a restored wagon, *Britannia,* with his fiancée and other friends as crew. The tour succeeded, although the wagon finished its journey in Florida, about 5,000 miles short of a complete circumnavigation, after all manner of hair-raising escapades. This journey has been widely reported in numerous publications so there is little need to cover it in depth here.

In January 1973, the citizens of the Common Market must have wondered who these crazy English people were that they had so rashly admitted to their Community, for the event was marked by a vintage and veteran vehicle rally to Brussels, organised jointly by the Veteran CC, the Vintage SCC, the Historic Commercial Vehicle Club, and the Vintage Motor Cycle Club. There were more than 100 vehicles in all, including ten old commercials, one of which was the author's Morris tonner prototype (now in the Leyland museum); another was a C-type Foden six-tonner belonging to Jack Hardwick of Ewell, Surrey, and fired by Bill Rivett, who himself owned a D-type tractor.

All went well as far as Folkestone, and the priceless contingent crossed to Ostend on a special ferry provided by Sealink. The Foden had to be towed 'cold' on to the ferry because of marine regulations but, as luck would have it, the chief engineer, himself an old steam engineer, insisted that the Foden be steamed up ready for disembarkation. The trouble was that the ship was a non-scheduled sailing, and everyone was kept waiting on the decidedly smoky car deck for nearly two hours while the waterfront bars were scoured for the man who operated the ship-to-shore ramp at Ostend. When he was finally found, and the vehicles were discharged, it was nearly 4 pm and almost dark.

The Bentleys and Invictas and the like simply roared off down the Ostend-Brussels Motorway and were in the capital within an hour, but the commercials and the slower cars had to go down the old road through Bruges, Ghent and Aalst, with its appalling *pavé* surfaces, because of the minimum speed requirement on the motorway. Most of the commercials kept in a little group headed by the author, who happened to know the way, and including Goff Radcliffe — another steam enthusiast — driving his Albion which had no lights at all, and with Vic Bignell, in a bullnose Morris complete with stop lights, bringing up the rear. The convoy was led off by the Foden, but at its first stop for water the steamer crew waved everyone by, and so were some way behind. After midnight had passed, the sponsors were just organising a search party with the aid of the Touring Secours people from the rally headquarters at the Europa Hotel, when a wisp of steam was seen rising past a window. Everyone rushed outside and there was the blackest wagon-crew you ever saw, grinning from ear to ear, having taken over eight hours to do the 70-mile trip from Ostend.

Trouble began, apparently, while still on board the ferry, when the long wait for the ramp operator meant that the fire had to be damped down in order to avoid asphyxiating the entire complement of passengers. Once on the road, the wet foggy air did little to boost a fire that was not made from best Welsh coal (simply because you cannot buy it these days), and the full steam pressure was slow in coming. Stopping for water, they discovered that there was plenty, but they would have to buy it. Suspecting that they were being 'taken for a ride' they moved on to the next source of water. By the time it became clear that they really were going to have to buy water, the level in the boiler was dangerously low and the fire had to be dropped. In getting things steamed up again quickly, they managed to set fire to the boiler lagging. They made their halting way to Brussels in the dark, arriving just before 1 am.

The next day things were different. All the motor vehicles trundled sedately down to the Grand Place for a public review, enjoying a very dignified and pleasant occasion. Until the steamer arrived, that is. As Jack tooted the whistle turning into the great mediaeval square with its gilded towers and mullioned windows, thousands of Bruxellois went wild with delight, and police had to force a way through the throng for the heavy wagon which was shining now in the morning sunlight after some hard overnight cleaning work. So tight was the schedule laid down for the rally that the spectators didn't get much chance to examine the masterpiece from Elworth; it had to set off right away in order to catch the boat back to England, which it managed with about 90 seconds to spare. But there's no doubt that the Belgians will remember that occasion and so in all probability will Jack Hardwick and Bill Rivett.

*Raking clinker out of the ashpan at a roadside stop on the 'Drive into Europe.'*

# Chapter 9

# Steam down below

Following the departure of Willy Foden and his family for Australia, new blood was brought into the works and new ideas were discussed. Many transport men in the industry had for a long time been declaring that the overtype engine on a locomotive boiler was too wasteful of chassis space, and that the boiler itself was too sensitive to water level problems in hilly country. The undertype, they said, was much better. What they probably meant was that the Sentinel wagon, with its vertical boiler and undertype engine, was a very good wagon, and why didn't Fodens copy that layout?

Sentinel were based at Shrewsbury, where they had moved from Glasgow in 1915, having formerly been known as Alley & McLellan, and there can be no doubt that Sentinel, as arch-exponents of the undertype wagon, made an exceedingly good product which was fast, quiet, economical, safe, and had a large loadspace compared with the Foden. Most of the major heavy haulage fleets ran Sentinels, often alongside the Fodens, and good though the Foden wagons were, there were many who reckoned that the Sentinel was better. Of course such a competitive situation was healthy for both concerns, and a constant succession of modifications emerged from both factories to improve their products. However, pressure was increasing at Fodens to boost flagging sales and finally, in the closing months of 1924, work began on a wagon that was entirely new to Fodens, something they had never contemplated before: an undertype wagon.

Fodens were, by then, practically the only producers of the overtype, all their major rivals having switched to undertypes many years before. It was, perhaps, inevitable that Fodens should finally go that way too, though there are still those who contend that it was a mistake and that the overtype should have been continued. Against that kind of argument and counter-argument, it is easy to see how controversial the decision to build an undertype must have been in the Elworth works, which was still relatively small, with about 550 employees.

The first undertype wagon was hardly the most handsome machine that Fodens had ever built. It lacked the businesslike appearance of the C-type, and was considerably heavier too. But in principle a lot of users liked it because it had a very large load platform, and the cab was all enclosed. The undertype engine was a two-cylinder non-compound, installed transversely across the frame below floor level, just behind the front axle. The crankshaft end was mounted slightly lower, giving the engine a sloping attitude in the chassis. This engine developed about 45 hp, which was considerably more than the C-type, and its boiler was mounted up front under the cab. The first boilers were plain cylindrical types

**Above right** *The first E-type wagon, with its upright boiler and transverse engine behind the front axle, was somewhat less reliable than the C-type with its loco boiler, despite the proclamations on the first exhibition machine.*

**Right** *In the E-type wagon, the two-cylinder compound engine was mounted transversely just behind the front axle, driving through shafts to the rear axle.*

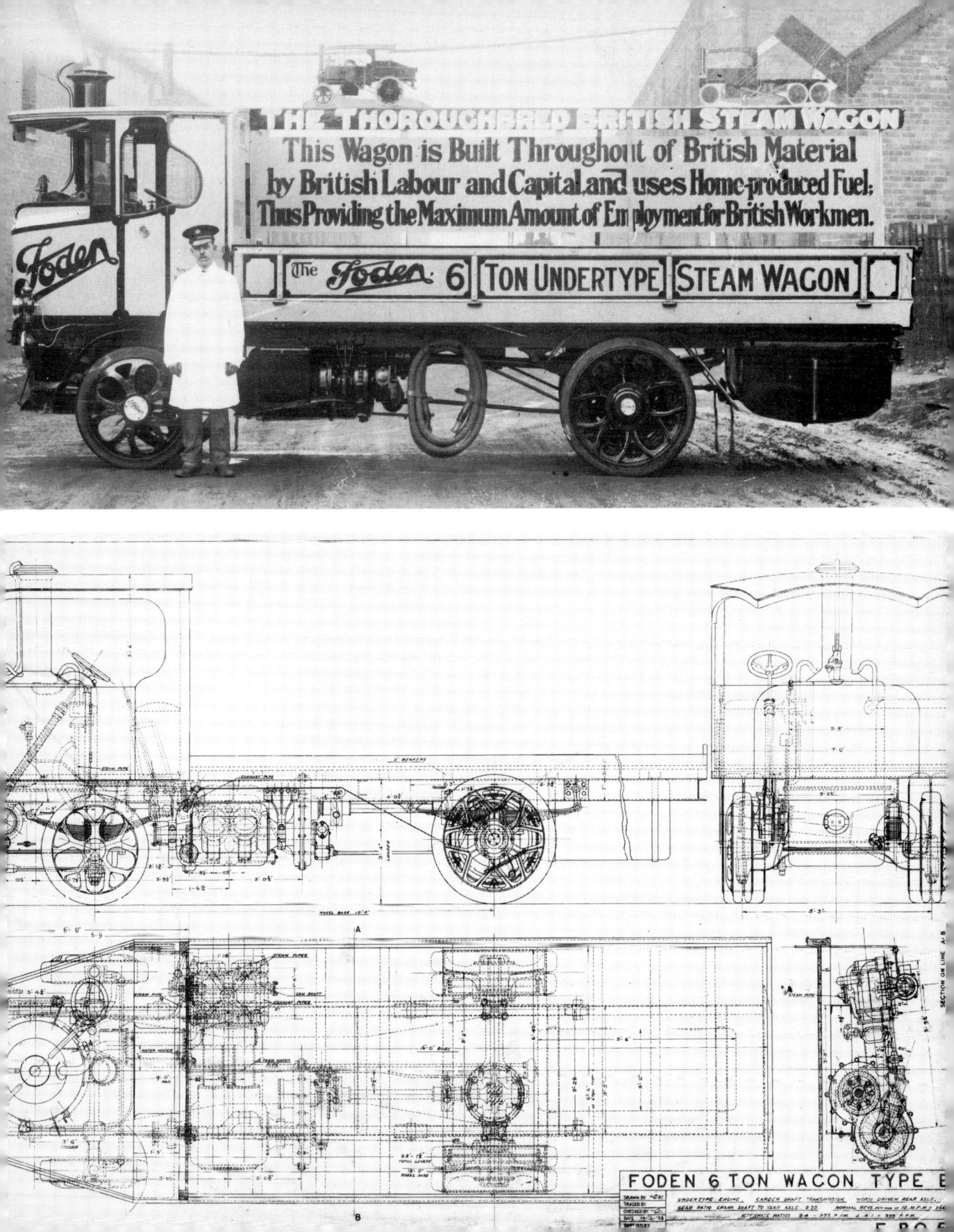

**THE THOROUGHBRED BRITISH STEAM WAGON**
This Wagon is Built Throughout of British Material by British Labour and Capital and uses Home-produced Fuel, Thus Providing the Maximum Amount of Employment for British Workmen.

The Foden 6 Ton Undertype Steam Wagon

FODEN 6 TON WAGON TYPE E

*Many Fodens were exported, to work on docks and industrial installations, such as this E-type discharging a ship in Buenos Aires.*

*Early Speed-Six wagons had the firebox protruding through the front panels of the cab, although later this was fully enclosed behind a rounded nose.*

with rectangular fireboxes underneath, heating the water through a complex criss-cross arrangement of tubes. The wagon was called the E-type, and it was the first Foden that needed a two-man crew. Two-man crews were recommended for the overtypes it is true, especially the faster ones, but it was quite possible for one man to do it on his own. In the E-type, the driver sat in the right-hand part of the cab, while his fireman was on the other side with the boiler in between them. The firebox was at that side, and it would have been quite impossible for the driver to reach the firebox, even had he wished to. Many E-types were built with a partition between the two halves of the cab, although the precise reason for this seems to be lost in obscurity now. It certainly did nothing to promote the team work that was essential for first-class steam-wagon operation. Having an engine of generous size, the E-type was considered to need only a two-speed gearbox, and this was installed next to the engine under the frame. It was a sliding-dog type box, a big improvement over the sliding gear mechanism employed on the overtypes, and it drove through a heavy shaft to the conventional worm-drive rear axle.

Early E-types were all produced on cast wheels with solid tyres, which contributed something to their peculiar appearance, as these wheels always looked too small for the rest of the wagon. However, various efforts to improve the wagon also improved its appearance, not least the moving forward of the front axle to a position under the extreme front of the boiler instead of under the centre of the firebox. This improved the weight distribution considerably, and made the wagon easier to handle, particularly on soft ground, although the turning circle was marginally enlarged as a result. There were numerous problems with the E-type, as might be expected when a switch is made to a new design concept after a quarter of a century's experience with a totally different concept, but one at a time the problems were tackled and solved.

Trade was still very poor in 1926. Orders would have been hard to come by even with fully proven products, and the general level of business slowly subsided. Efforts were made to boost trade by export business, but that was only partly successful: a series of tractor versions of the E-type was made, some for South Africa, some for Argentina, and odd ones for other parts of

*Steam down below*

the world. But, useful though such orders were, they did little to boost the overall fortunes of the company and, though things were still profitable, just, and most of the workforce had been retained, morale was low, and little confidence was enjoyed between the workforce and the new managers who had been brought in after the family's departure for Australia a couple of years before.

Fodens even ventured into the railway business at one point which proved, if nothing else, the versatility of the E-type design. It was in effect an E-type tractor on a short wheelbase, like the export models but fitted with fixed axles and flanged wheels suitable for running on the railways. This vehicle, subsequently used as a shunting locomotive by the Palmer Mann Salt Co, was built in a rather remote corner of the works, and had to be manhandled on wooden skids about 200 yards to the railway sidings at the foundry end of the works; this move included at least four 90° changes of direction. As far as can be ascertained this was the only E-type railway locomotive, although an experimental passenger railcar was built, using E-type mechanical units. Sadly neither of these ventures bore fruit commercially.

Despite the problems with the E-type, there was a certain amount of encouragement from users who favoured the undertype, and so an improved wagon was put in hand in the design office. This was known as the 'Speed-Six', or O-type wagon, so called because it had a six-ton basic chassis capacity. There may be some truth too in the alleged notion that the Bentley Company was then enjoying considerable success with its performance cars, one of which was called the 'Speed-Six', and there might have been some little advantage to be gained by sharing a famous name.

In many respects, the production 'Speed-Six' Foden was a remarkable machine when it emerged in 1928. It used a developed version of the same two-cylinder transverse engine that had powered the E-type, but had a more elaborate transmission with three speeds, 40 × 8 pneumatic tyres all round, a fully enclosed saloon cab, and a very high performance. The works development 'Speed-Six' chassis could do almost 60 mph fully loaded, which must have been quite exciting in days when maximum speed limits for lorries were still in the 12 mph bracket, or 20 mph if you met certain weight conditions. In fact production machines were set up to do about 45 mph, but of course many of them were 'adjusted' by their owners to do higher speeds.

The early 'Sixes' suffered from a firebox

*The Speed-Six wagon developed from the E-type was powered by a transverse two-cylinder compound engine, and was capable of something like 56-60 mph in its later versions.*

82                                                                                          The Foden Story

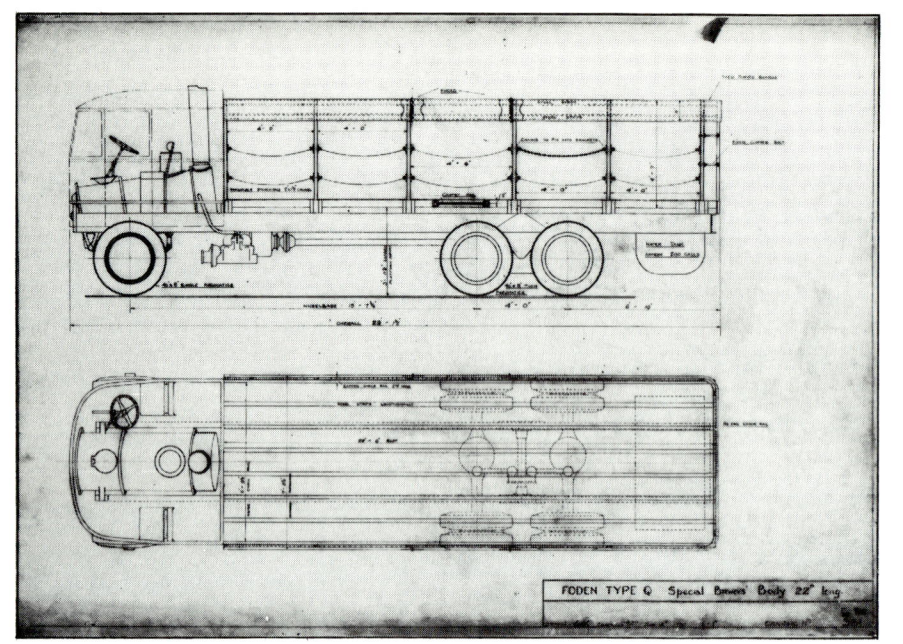

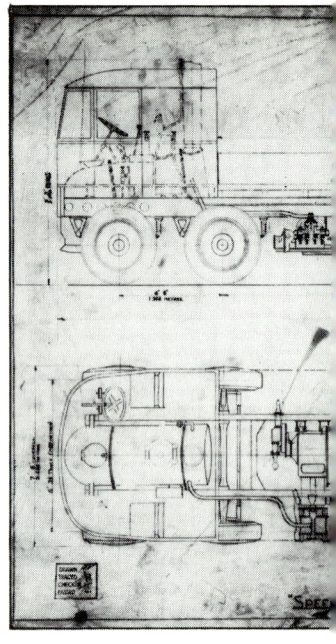

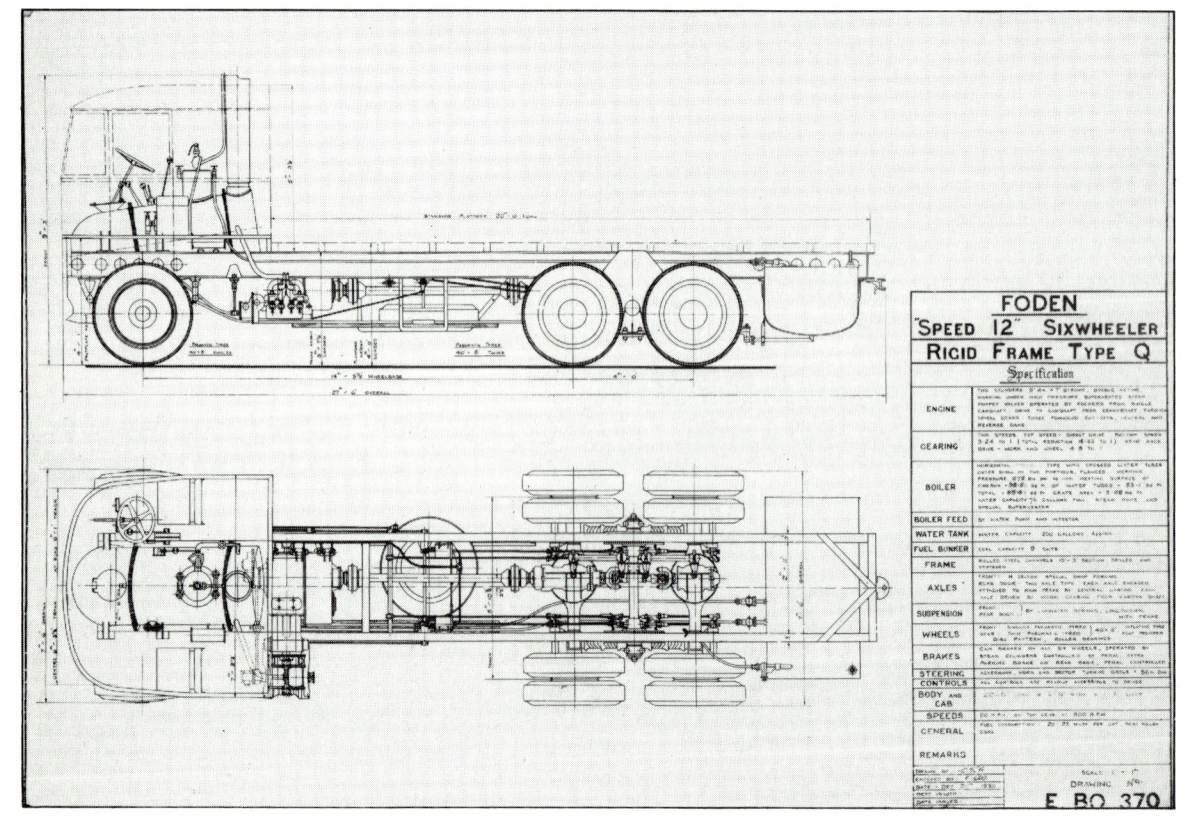

*Steam down below*

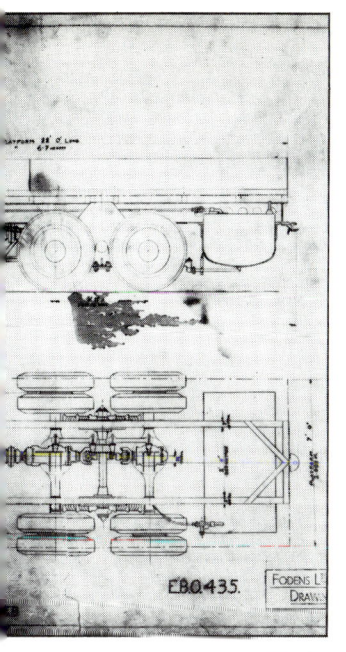

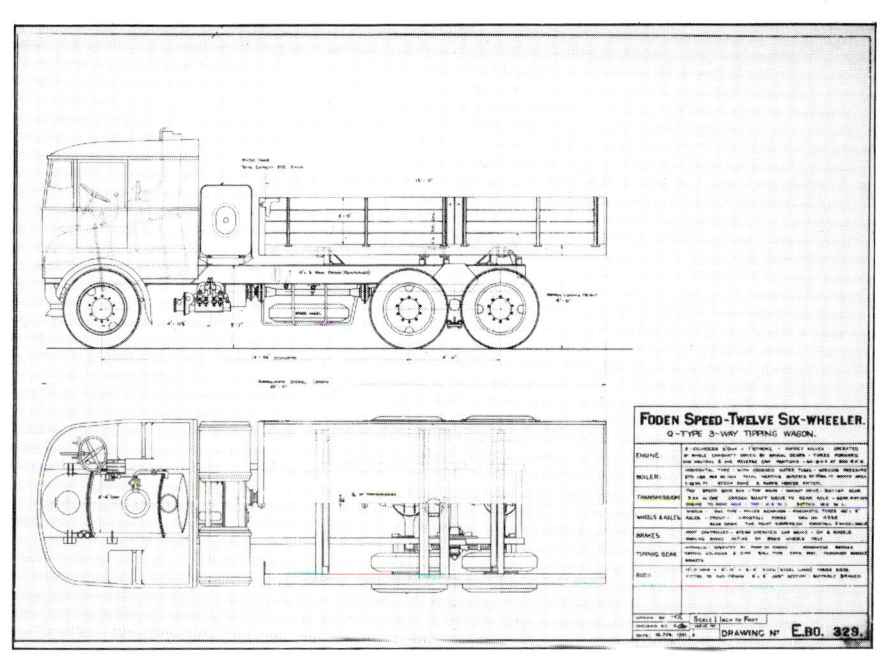

*Various versions of the Speed-Twelve were offered but, although designs were produced in 1931, the four-axle version was never built.*

problem, and were inclined to get a bit breathless if used at speed for any length of time, and before long a modified firebox and boiler arrangement became necessary. This took the form of a sloping cylindrical boiler, downturned at the front, giving it the nickname the 'banana boiler' at the works. It had a forced-air system under the front of the chassis, and could be fired from on top, as distinct from the side-firing method of the E-type. This boiler was of all-welded construction, a fact which did not go down too well with the boilermakers who had been at the heart of Fodens' activity for 50 years, and who now found themselves and their traditional riveting methods unwanted. Just as there were six-wheeler versions of the six-ton C-type, so six-wheeler versions of the undertypes were built. Not illogically, the 'Speed-Twelve' label was applied to the stretched 'Speed-Six', and these machines won a certain amount of approval among operators for their capacity for shifting freight over long distances at unprecedented speeds.

The sight of a steam wagon travelling at speeds in the 40-50 mph bracket or higher was particularly impressive, perhaps because one tended to associate this kind of machine with a more sedate pace altogether. Nevertheless, anyone who has seen a Sentinel or a Foden undertype with a full load, bowling along in almost complete silence at that kind of speed is never likely to forget it and, to many steam men, those brief years at the end of the '20s and the beginning of the '30s represented the pinnacle of achievement in the steam wagon industry. Unfortunately, despite the impressive nature of those latest undertypes from Elworth, and the undoubted effectiveness of them as freight carriers, their days were numbered. The internal-combustion-engined vehicle was fast ousting the steamer as the mainstay of road transport, as users demanded more flexibility, less time-wasting in the mornings, cleaner working conditions, and greater versatility.

At Elworth the board appointed a succession of general managers in attempts to grapple with the enormous tasks and decisions facing the company. E. R. Foden was the only family member remaining, and he found himself in a position of diminishing power as less experienced members overruled him on all manner of matters. Not least among these was the way in which

**Left** *Speed-Twelve wagons were relatively popular with bulk hauliers, but steamers were rapidly losing ground by then to internal-combustion-engined lorries. This picture was taken in 1931, the same year that Foden built their first diesel.*

**Below and facing page** *The Speed-Six was a versatile chassis. However it is doubtful if all these versions were built. There is no evidence, for example, of the existence of a Speed-Six bus.*

long-serving employees were dismissed without notice should a temporary lack of work be discovered in their particular department. Such procedures were very painful to a man who had been with the company all his life, and who had seen it grow from almost literally the ground upwards. Many of the older employees had been there as long as he had, yet their loyalties seemingly counted for nothing.

As the new decade of the 1930s stumbled on, further schemes to revive some trading interest were mounted. There were two further steam-wagon designs, the N- and the Q-types, and numerous other ideas were tried, in particular the use of the basic undertype engine in agricultural tractors, but mounted in the overtype position atop the boiler. Two of these were developed, the 'Agritractor' and the 'Sun' tractor and, though both were very powerful and reliable, there were very few parts of the world where a steam tractor could be sold, because the powerful and simple kerosene tractors from American farmlands were taking over agriculture in an overwhelming fashion. Also, even in 1928-30, these Foden tractors looked old-fashioned with their wire-spoked wheels and straked-iron tyres. Consequently they made little contribution to the survival of the company, although a small

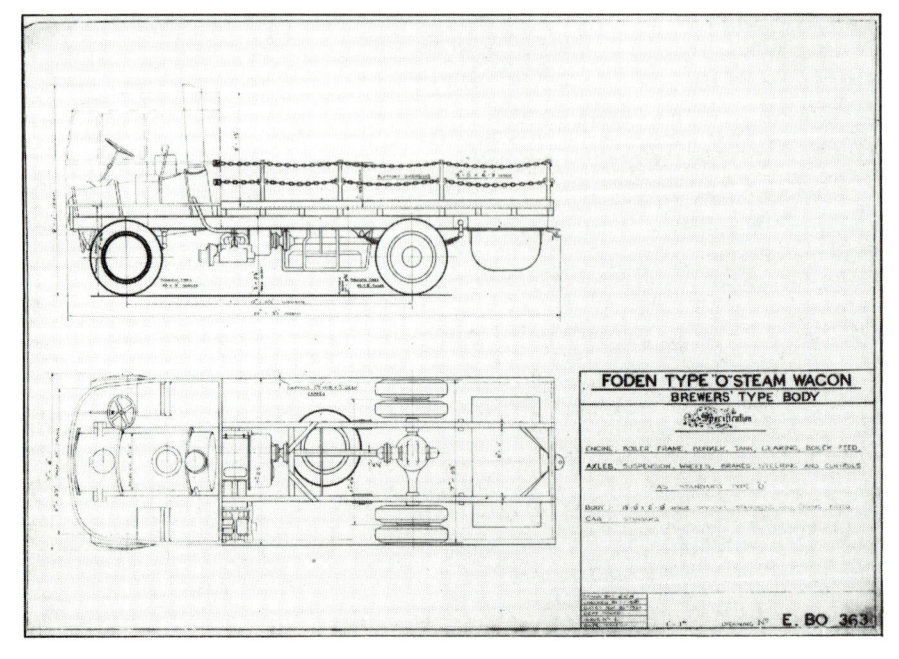

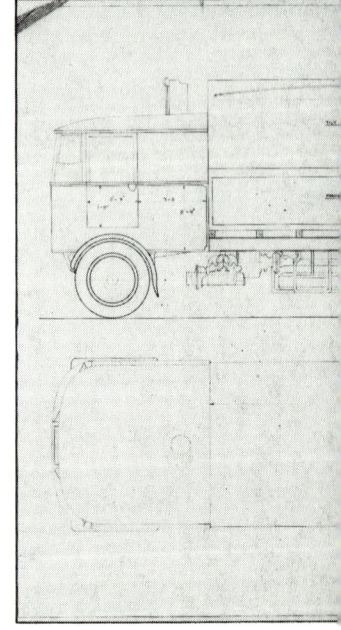

*Steam down below*

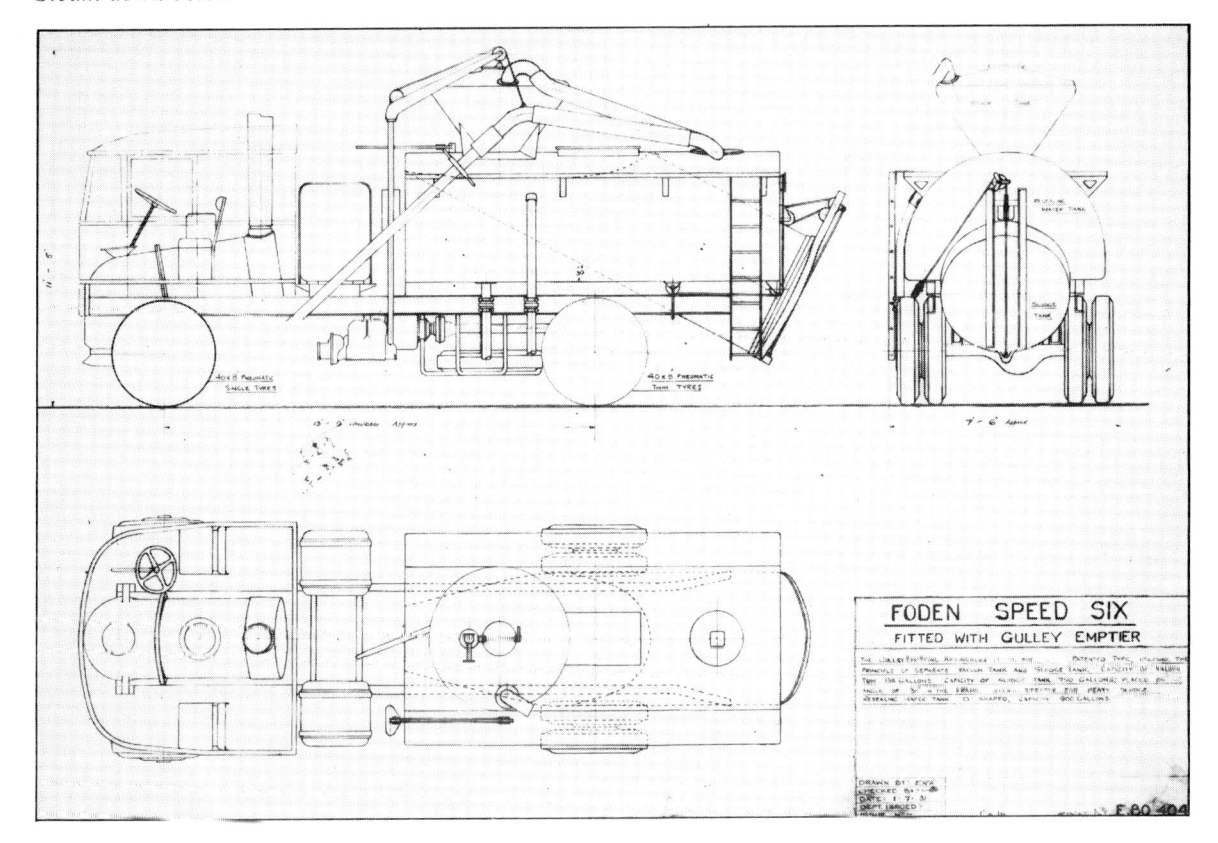

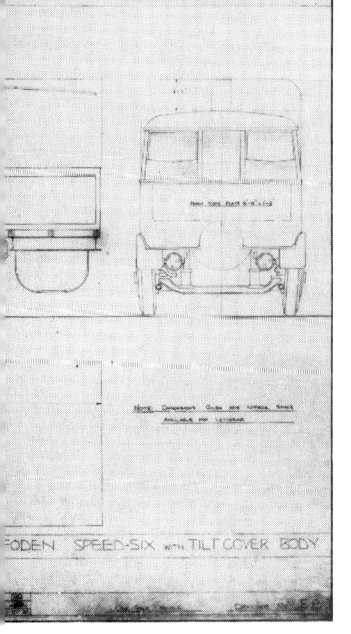

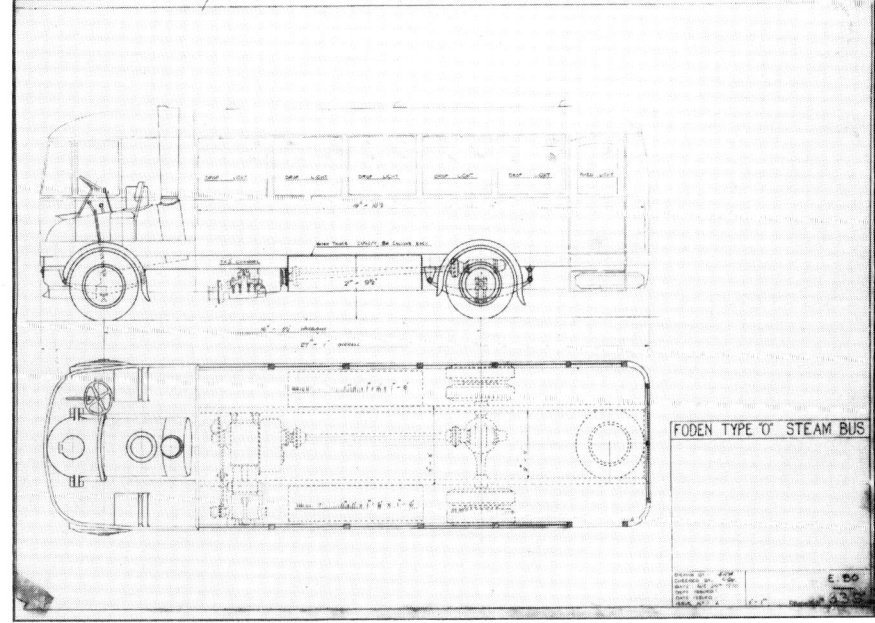

number of road-going 'Sun' tractors were built with solid-rubber wagon-type wheels for use on docks and freight yards, where they played a useful part for several years.

The 'Q'-type wagon was, in effect, an experimental higher-power Speed-Twelve with yet another boiler arrangement, designed to need less attention on the road than the usual type, which required close supervision by the fireman if succeeding cycles of excess steam and insufficient steam were to be avoided. In this it was partly successful — but it never went into proper production, so disorganised had the works become at that stage.

The N-type wagon was a 12-ton six-wheeler which never got beyond the prototype stage. Not that it didn't have a lot of clever features; it did, but by that time it was too late, and the steam wagon was yesterday's machine. Its engine was again a two-cylinder compound with poppet valves instead of slide valves, and it drove to a three-speed sliding-dog type gearbox with what would now be called a conventional gear lever. Instead of a normal regulator lever, the speed was controlled by a pedal in the same position as one could find the throttle pedal on a petrol-engined lorry. The boiler was mounted vertically to the left of the driver and behind him, so that the fireman travelled with his back to the direction of travel. The N was fitted with hydraulic brakes, the fluid for which Fodens manufactured themselves by mixing castor oil and alcohol! Only the rear axles carried brakes, with aluminium shoes inside large cast-iron drums, and a unique feature was that a hand valve could select brakes on either axle alone, or on both axles together. There is no doubt that the new wagon's performance was impressive and it is claimed that the 'N' was the first wagon to climb the old Mam Tor road in north Derbyshire, then steeper than 1 in 4, with a full 12-ton load. On one occasion, when descending that same road, a brake shoe disintegrated and smashed a wheel cylinder, allowing the fluid to escape. The driver had the presence of mind to switch immediately to the other axle's brake system and descended safely. That wagon was built in 1932-33 and, despite its undoubted ability, it was never a commercially successful machine. One stalwart operator struggled on with it, working around Derbyshire, and at one stage its Foden boiler was removed and replaced by a Sentinel boiler. But that was the last of the Foden steamers; a new age had dawned.

*The long downswept boiler arrangement and difficult driving conditions of the Q-type wagon limited its popularity with crews, despite a promising performance.*

# Chapter 10

# Not so much a job, more a way of life

When, at the turn of the century, the success of Edwin Foden's steam wagon demanded that an expansion programme be undertaken at Elworth, one of the major problems was recruiting the necessary skilled labour. There is no doubt that a very high degree of skill was needed. The design of the Foden compound engine, for example, was such that it needed accurate casting and machining, and equally accurate assembly. Likewise the boilers needed first-class workmanship if they were to operate safely at pressures of some 200 psi. In addition there was a great quantity of fabrication to be done in iron, steel and brass. The rule that governed all this work was that weight should be kept to an absolute minimum. This meant not only accuracy but consistency, since speed limits hinged round an unladen weight not exceeding three tons. It so happened that steam wagons did not have to include coal or water in that weight, and an unladen road-going wagon was probably nearer four and a half tons but, as long as it came under three tons 'dry', that was satisfactory as far as the law was concerned.

Consequently, when recruitment began for the new and enlarged works in 1902, Edwin and his two sons William and E.R. concentrated on finding men with the necessary skills, while their new fellow-directors, led by Cecil Brunner, looked after the financial side. It was undoubtedly this combination of attention to both engineering matters and financial soundness which put Fodens Ltd on the secure footing that saw it through to unprecedented prosperity in later years. Too many firms, particularly in engineering, relied heavily on their engineering prowess, and insufficient attention was paid to commercial matters with the result that, despite a high-quality product, they went out of business. This did not happen at Fodens Ltd. So strong was the company in those steam days that the high profit levels of almost a quarter of a million pounds achieved just after World War 1, and before the depression, were not reached again until 1951.

There was no shortage of would-be workers when the word went out that Fodens were expanding. The little Elworth company had made an enviable record for itself with successful agricultural machines, great industrial engines and new steam wagons. The Victorian paternalism of Edwin Foden was well known and, although the employees had to put in an exceedingly hard day's work, the rewards were good and the security of employment better than in most places. Furthermore, there was a welfare department and a chance of help with housing, if not actually a company house. In any concern, expanding the workforce from about 50 men to over 200 in a few months, and then continuing to expand it in succeeding years must be a delicate process; indeed in many it could be disastrous. But Fodens accomplished it with a minimum of turmoil, and an astonishingly smooth transition took place from what was essentially a highly specialised engineering shop where products were largely 'one-off' designs, to a series-production factory where parts and complete vehicles were built to very close tolerances in considerable numbers. From the time Edwin and William

**Left** *In the board room, which doubled as the founder's office, a stern-faced Edwin Foden kept an eye on company business. This picture was probably taken in about 1910.*

**Right** *The 1902-03 office building on the left is opposite the foundry (right), itself on the site of the old Elworth Foundry. The works was gradually extended down both sides of Elworth Road (centre) in future years.*

drove their tough little steamer back from Hampshire at Christmas 1901, to the commencement of series-production of wagons with a newly recruited labour force, took less than 12 months, and that period included the design and development of improved features in the wagon itself.

In due course the works expansion included a modernised foundry, a new boiler shop and a new erecting shop where the wagons were assembled. These new buildings extended down both sides of Elworth Road, where the original foundry stood, and where much of the works still remains. The factory was, for its time, a large one, and the problems of supervision were greater than the Fodens had experienced before. But they took care to ensure that they knew every man who worked there, what his particular skills were, and what his family was like. To be sure the ways of the pre-wagon days had to change. No longer could Edwin or his sons chat as man to man with each and every employee, although relationships based on respect of skills and loyalty remained as a strong force within the company.

New employees soon learned that there could be no short cuts in their work. For Edwin Foden and his two sons there was only one way to do a job, and that was to do it properly. And if the job was not done properly the awesome prospect loomed of a visit from the boss. Few risked the humiliation of being shown by a suited and hatted employer exactly how a bearing should be scraped or a boiler-seam caulked, in front of his fellow workers. It was this formidable knowledge that the boss could do any of the men's jobs that kept up the standards.

Few jobs had the same attraction as working at Fodens Ltd, for at up to 30 shillings a week for skilled men the pay was good. If you earned any degree of seniority you would probably be able to rent a very good house from the company at a suitably low cost, and the welfare department would look after cases of sickness or industrial injury, long before the days when such things were required by law. So as the first decade of the 20th century passed, the new and enlarged company was rapidly knitting itself into a large industrial family, and not simply an anonymous collection of artisans working for a domineering employer.

When Edwin Foden died in August 1911, many thought that things might change for, whereas he was undoubtedly a Victorian in all the best meanings of the word, E.R. and William had more advanced thoughts on the way things should be done. In the event, things did change, but not in the way many had expected. From

1911 onwards there was a rapid acceleration in the building of new workshops, installing of new machinery, and the building of new houses for the workforce. The factory investments are dealt with more fully elsewhere, but the housing investments warrant a special examination.

At the time Edwin died there were about 40 houses and cottages in company ownership, including the founder's house which is now part of the office accommodation on the corner of Hill Street in Elworth. The Foden family occupied some of these houses, while senior men like Fred Mason who had been with the company for many years occupied others. After the expansion under the new company name, building and purchasing of cottages for employees continued steadily, although transporting workers from an increasingly wide area raised problems. Many walked several miles, or cycled even greater distances. Public transport was very limited, and the operation of a works bus was too expensive. Clearly there was an urgent need to house the men and their families closer to their place of work. In 1912 20 new houses were acquired or built, all within half a mile of the works and, in the following year, a whole street of about 50 houses was built. This was George Street, approximately 200 yards from the works, which still stands today as a monument to the community that the Fodens built in Elworth.

Those houses were no ordinary workers' cottages: very substantial construction was used, and the walls were finished in the highest-quality facing brick, in exactly the same manner as the directors' houses that were built at about the same time. They had proper kitchens and bathrooms, a small garden in front, and a larger garden at the back where vegetables could be grown if the family was so inclined. There were also allotments not far away where strips of good growing land could be rented for a nominal sum, for the cultivation of fruit and vegetables. Looking at George Street today it is hard to appreciate just how long ago it was built. The neat houses look very much less than 65 years old, especially those where modern windows have replaced the original sash frames.

On the land between George Street and the factory, Fodens Ltd established their 'Recreation Club'. This originally consisted of a bowling green and some pleasant gardens, with a small building where indoor leisure activities took place. It was established before Edwin died, and over the years expanded considerably, eventually outgrowing the available space on the site; a

*The founder's house, now somewhat overshadowed by the office headquarters erected in 1956, was an elegant building in the best Victorian tradition.*

new and larger sports facility was therefore set up in Sandbach itself, with football fields, hockey pitches and changing rooms. However the headquarters of recreational activities has always been at the original site near the works and, indeed, for many years the annual general meetings of shareholders were held in the club buildings. The running and maintenance of the club facilities is financed by the employees themselves, and has been since the early days. A penny was deducted from the wage packets each week, and a small additional sum was added by the company, mainly to pay the staff needed to look after the facilities. Sales of drinks and food in the club, at prices well below those available elsewhere, also made a good profit and contributed to the success of the operation. Even today, when company loyalty and community spirit are mere shadows of what they were in the 1930s, the leisure facilities at Fodens Ltd contribute significantly to the loyalty and dedication of the employees, and over 90 per cent of the current workforce of nearly 3,000 holds membership of the club.

Back in 1912 such facilities were unusual, to say the least. There were some towns where benevolent people had donated land and funds for a park or sports field, but to have such wide-ranging facilities within a relatively small company with, at that time, no more than 400 employees was exceptional in industrial Britain. The policy certainly paid off. At any one time there was a waiting list of men who wanted to work at Fodens, and consequently the company had the pick of the best artisans in the area.

When the 1914-18 war began, few believed it would last for long, but nevertheless there was a sense of urgency among industrial communities in the Midlands and north-west of England, that did not seem to gain strength in areas such as the Home Counties until much later in the war. From the outset, Fodens were turning out heavy artillery shells and steam wagons. The wagons were built both for the War Office and for civilian users working on military and essential 'home-front' work. The average working day went up from ten to 12 hours, and for the first time a night-work shift was started. Female labour was used for the first time too, although limited in quantity. Even the band, by then an almost sacred institution, gave up full-time music and came to work on the factory floor, practising music in their few remaining leisure hours.

There can be no argument that the war was

Not so much a job, more a way of life

financially of great benefit to Fodens, but the earnings did not all go into shareholders' pockets. Immediately after the war was over, further housing developments took place, so that more employees could enjoy the benefit of low-cost company housing. The Recreation Club was further expanded with a second bowling green and a big new club building with a bar and other social facilities. In the works medical facilities were improved with the provision of a proper nursing attendant and a fully equipped sick-bay.

Inevitably, as the depression of the '20s gripped the economy of Britain harder and harder, work conditions suffered. Wages were cut several times in the early and mid-'20s, and there was a steady reduction in the number of employees. Until William Foden went to Australia in 1924 the cutbacks were restricted to the minimum necessary, but once E.R. was on his own as a family director he found himself outnumbered on the board, and far more savage lay-offs and wage cuts were made. This was undoubtedly a black period in the history of social relations at Elworth, and a great number of cases of hardship and suffering can be traced, where houses were sold off in order to bolster the flagging fortunes of the company, while the occupants found themselves out on the street at

**Above** *An entire street — George Street — was among the property built by Fodens before World War 1 to house employees close to the factory. In 1978 these houses still look surprisingly modern.*

**Below and overleaf** *This series of pictures shows the interior of the Elworth works at the height of its prosperity in the early '20s. All castings from small valves to wheels were made in the foundry.*

*Iron and steel parts were hammered into shape in the forge.*

*All manner of cast, forged and fabricated parts were finished in the machine shop.*

*Fireboxes, boilers and smokeboxes were made in the boiler shop.*

*All the components were put together in the erecting shop, across the road from the foundry and boiler shop.*

short notice, as often as not without a job. Despite all this a degree of loyalty remained, as demonstrated during the General Strike in 1926 when all the Foden employees stayed at work instead of following the dictates of labour leaders. Their action was probably prompted as much by fear of losing their job as by loyalty to the company, but the fact remains that they turned up to work when practically the whole of Britain was on strike.

Working conditions in the factory were not, by that time, as good as they should have been. Although, at the turn of the century, the new works represented a high degree of improvement on the conditions generally found in the engineering industry, it had not really kept pace with the times, and by the mid-'20s conditions were far from ideal.

The men in the boiler shop and erection shops often needed extra light when working on the inside of boilers and tanks, and candles were the only available source. Usually the candle would be wedged into a suitable-sized nut, of which there were always plenty lying around, but in some cases a short stubby candle would be fixed to the peak of a man's cap. Much of the work was unpleasant in itself. Foundries are always rather dirty places to work, even the mechanised ones of the 1970s, but 50 years ago Fodens' foundry was not of the best, particularly as increased production over the years had led to a piecemeal extension of the place, with too many corners and odd-shaped areas, inadequately lit and ventilated.

The new boiler shop was better with its high roof and good natural lighting, but the noise level with all that riveting and caulking going on was absolutely shattering, and partial deafness was a common problem among those who worked there. Even the erecting shop was not without its difficulties. One of the less pleasant jobs was painting the inside of the water tanks on the wagons with a red lead paint, a task usually given to the apprentices. They had to crawl in through the tank manhole and carefully paint the whole interior with red lead. Inevitably this meant that they ended up with a lot of paint on their clothes and skin, so it was a messy business. It was normal practice to have a second apprentice bang loudly on the side of the tank every couple of minutes in case the boy working inside should have been overcome by the paint fumes. But despite these conditions, few felt that the situation warranted any complaint.

Some improvement came in 1927 with the installation of a further diesel generating-set and the extension of mains electricity, so that all the shops were lit by powerful electric lamps, and the inside work on boilers and the like could be done with the aid of hand electric lamps. But even in the 1930s heating consisted merely of two hot-water pipes running through each workshop, fed by ancient coal stoves. It was quite usual, especially in winter time, for the men to stop work periodically and beat their arms about their chests to generate a little warmth in their hands and bodies so that they could keep the feel of the tools and wagon-parts which relied so much on their manual skills.

It should not be forgotten that it was very heavy work where steam wagons were concerned. Despite the surprisingly low weight of a complete wagon, there were not many parts, and individually they were very heavy to work with. A rear axle, for example, demanded Herculean strength to get it in position under the frame, and even a cast-steel wheel with its solid rubber tyre was as much as one man could handle on his own. For most of the period of steam-wagon production little in the way of mechanical handling aids was available, beyond a few chain blocks for lifting the very heavy sub-assemblies such as completed boilers and chassis frames. All the rest was manual work and, ironically, it was only when the much lighter diesel trucks started coming on stream that any significant number of electric and hydraulic hoists were available for handling the products.

A normal working day in the 1920s and early '30s started at 6 am, which meant that for much of the year a man went to work in the dark. He would work until 8.30 when there was a ten-minute break for breakfast. This he would usually bring with him from home, and it invariably consisted of bread with perhaps some cheese if he was lucky. The youngest apprentices had the responsibility of fetching 'the brew': Each man had a 'brew can', which was a metal or enamelled canister with a lid. A quantity of tea would be placed in the can, and the apprentices took all the cans from their particular shop to a small boiler which dispensed hot water. These trays of cans would be ready just as breakfast began, and the can lid was used as a cup to drink the precious brew. The morning's work continued until 12.30 when an hour's break for the midday meal took place. Some men went home for that meal if they lived close to the factory, while others had their own snack, or 'snap' as it was usually called, with them. Again the brew trays would be pressed into service. Afterwards, work would continue right through until 5.30 or sometimes 6 pm if there was plenty to do. Saturday mornings followed the same pattern as a weekday, except that at noon all work ceased until the following Monday morning.

An apprentice was paid 4s 6d in his first year, and that was increased by 2s a year for five or six years, depending on the terms of his apprenticeship. At the age of 21 he was considered to be a skilled craftsman, and was usually taken on to the main workforce, but he would not receive a full wage until he was 23, and then his weekly reward would be between £2 10s and £2 16s, according to his trade.

The young apprentices had another job to do on payday, and that was to sort out all the numbered brass containers that were used in the paying-out process. These brass cups were kept in a big basket in the pay office (by the gate just off the Elworth Road, which ran through the factory). The cups were arranged in numerical order in wooden trays, and the appropriate sum of money was placed by the wages clerk in the cup bearing each man's works number. As each man collected his money, he would check it and,

*Pay day at Elworth in 1940.*

*Employees' pay was placed in small brass containers bearing their works number.*

when satisfied, he would drop the brass container back in the big basket.

Office staff had equally difficult times. Invariably there would be a bargaining session over wages before their engagement was agreed. For example Clifford Brassington, who became chief buyer for the company, recalled the day he went to Fodens for a job in 1933. He asked for 27s 6d a week, but was offered £1 or the job would go to another applicant. A little later he was threatened with the sack for using a quarto size sheet of paper for a five-line memo. Waste was not tolerated at Elworth. On another occasion, Mr Willy walked through the stores-receiving building without warning and found too many people apparently doing nothing. Mr Brassington was in the direct line of sight and Mr Willy said, 'Too many men about in here, you'll have to go.' A moment later he turned round to say, 'Revise that: you stay, I'll sack someone else instead.' Such peremptory handling of labour relations would not be tolerated in this day and age, but 40 to 50 years ago it was normal, and nobody grumbled.

In the works there was one special breed of men — the drivers. Their job was to deliver wagons all over the country and to test them after they had been built or had been back to the factory for repair or overhaul; they were undoubtedly the 'kings' among the employees. Even senior engineers and supervisors would not dare drive a wagon, although they would sometimes be allowed to fire while a driver drove the machine, particularly in the later days when the undertype wagons needed a two-man crew. But the overtypes could be handled by one man alone, and this was the period when the drivers acquired their exalted rank. To some extent this arrangement lives on, and even in the 1970s the works drivers are a breed apart with a seniority by tradition rather than by definition over their fellow-workers.

During the worst phase of Fodens' affairs in the early '30s, the workforce was heavily cut back, and those who retained their jobs were frequently required to work one week in two, or sometimes as little as one week in four; this, of course, meant only one wage in four. Furthermore the basic wage for skilled men was cut back, sinking as low as £1 12s 6d at one stage, when even the repair shop was empty because there was so little money available among the transport users. Those were bad times indeed. When the Jarrow March to London took place in 1933, with the object of protesting to Parliament about the appalling unemployment and poverty in industrial areas, a sizeable contingent went from Sandbach and Elworth to join the main stream from Tyneside. To some extent they were considered heroes, fighting for a better living for themselves and their fellows and, when they set off, good wishes went with them from many of the local townsfolk and business people, and the town policemen escorted them on their way and wished them success. The reason was simple: everyone depended on local industries like Fodens. If there was no work at Fodens, then the village butcher wouldn't sell his meat; all kinds of livelihoods were threatened by the chain reaction.

Fortunes took a turn for the better as E.R. Foden's new company, ERF Ltd, began to thrive in Sandbach. Many men who had been dismissed from Fodens Ltd found work at the new factory up the road, small though it was. It signalled

**Right** *Long-service certificates presented to selected employees carried a pictorial history of the company. This certificate was presented to Harry Smith in June 1959.*

# Fodens Limited
## Elworth Works, Sandbach

# Long Service Certificate

PRESENTED BY
THE BOARD OF DIRECTORS TO

*W. H. Smith*

UPON THE OCCASION OF COMPLETING

*Thirty-one Years*

LOYAL AND FAITHFUL SERVICE
AND AS A MARK OF APPRECIATION
AND GOODWILL

William Foden
Governing Director

Date June 1959

better times ahead, and slowly the old firm followed the new one to prosperity. The products were by then completely different; diesel lorries. The new machines made a lot of difference to working conditions. The work was a lot easier because components were smaller and lighter, and by then more mechanical aids were available. On the debit side, there was no work for the boilermakers with their riveting and heavy plate-working skills; indeed, since the introduction of welded boilers around 1930, theirs had been a dying trade at Elworth. Some of them worked as steel fabricators — there was plenty of work of that kind on the diesels — making all the different items of chassis hardware from thick steel plate.

When the first Gardner engine arrived at the works, everyone gathered round to examine it. Few of them had ever seen a diesel engine, although some were familiar with petrol engines and, on that Wednesday afternoon in 1931, it seemed a strange shiny object, with little in common with the type of engineering to which they were accustomed. The L2 Gardner was basically a marine engine, although the version sent to Fodens was stripped of its marine equipment such as the sea pump and heat exchanger. Nevertheless it was an unfamiliar piece of machinery, and the first jobs were working out how to install it in a vehicle, and how it should be controlled. Quite obviously there were a lot of new skills to be learned before the newfangled power source could be mastered, and there were many who went home that evening convinced that they would be better off sticking to steam power, rather than becoming involved with the mysteries of the new engine. They were wrong, however, and the wiser hands in the factory faced the fact that, unless they learned all about diesels and how to handle them, they would be left behind by the tide of progress.

There were to be many trials in the following years with nearly as many errors and, under a management which after the end of 1932 contained no Fodens at all, but only outsiders who had limited experience of the problems involved in such a factory, there were to be many more dismissals, and many more heartbreaks before matters improved. In fact the situation deteriorated so much that a group of senior employees, supervisors, and others who had been with the company many years, joined together to send a telegram to Mr William Foden, asking him to come back and retrieve the situation. Mr Willy had been a hard boss to work for, but he had been scrupulously fair, and he understood the difference between a real problem and an imaginary one. In fact Mr Willy did come back, and that phase in the company's history — when it rose like a phoenix from the ashes of the early '30s — is dealt with in Chapter 14.

In more recent times, things have changed markedly. Few of the company houses are retained, because of the incompatibility of engineering manufacture with property dealing, and in any case most workers feel that they need a certain amount of independence from their employer and seek their own housing instead. Conditions in the new chassis-assembly plant are totally removed from the dark and dirty shops of the '20s, with spotlessly clean floors, controlled air temperatures, plenty of space and light, and a low noise level. Some of the older shops remain, but even these are much better than when the present employees' fathers worked there.

Despite the great social changes, loyalty is still an important factor in the company's success. There are still many men who have worked there all their lives, men like Albert Bourne who began in the early '20s as an apprentice and who, although now officially retired, still goes into the works one or two days a week to keep an eye on the two steamers and diesel number one which are kept at the works. Harry Smith is another who retired recently after more than 50 years with the company, and there are many men who work at Fodens now who would never think of working anywhere else, and who are following in their fathers' footsteps in devoting the whole of their working lives to Fodens. Despite the introduction of computer control for wages, all accountancy functions, and the handling of parts and components, there remains an indefinable personal aura about Fodens Ltd which undeniably stems from the long traditions of the company. That such traditions can live on in a highly technological age, is in itself something of a miracle. But then near-miracles are nothing unusual at Elworth. They've been working them for years.

# Chapter 11

# The Foden Motor Works Band

Foden steam wagons and diesel trucks became world-famous in the transport sectors of industry, and Foden marine engines became well-known in both civilian and military applications. But it was in an entirely different sphere that the name Foden became familiar to millions of ordinary people, to whom the workings of a steam or diesel engine were a complete mystery. That other sphere was music, or to be more precise, brass-band music.

For more than 75 years the Foden Motor Works Band has been a popular and highly skilled group, reaching the highest standards possible in the world of musical entertainment and competition. Yet the production of all that splendid music was rooted in a somewhat embarrassing occasion when there was no music at all.

The worthy inhabitants of Elworth were no strangers to the excitement of a celebration, no matter what the cause. Every possible state occasion was celebrated by a procession or fete, during which the local musicians, amateurs though they were, lent their utmost enthusiasm for the price of a pint or two of ale. Yet that ale was the cause of a serious crisis in the musical life of Elworth.

The occasion was the relief during the Boer War of the beleaguered town of Mafeking, where the forces of Lt-General Robert Baden-Powell had been under siege from the Boers for an uncomfortably long and politically embarrassing time. Finally, on May 17 1900, Mafeking was relieved by General Mahon's troops, and the occasion was treated by the whole nation, Elworth included, as a national triumph. The Elworth celebrations were well organised with processions of make-believe 'armies' and effigies of the Boer leaders, all headed somewhat incongruously by a new Foden traction engine. At nearby Sandbach, on the other hand, the council declared that the common should be the venue for celebrations, but no organised activity was forthcoming. Accordingly the Elworth contingent were asked to take their procession to Sandbach, to which they agreed on condition that they could borrow the Sandbach town band for the event. It was further agreed to detour through the small village of Wheelock where the village Temperance Society had a band, which was recruited for the day in order to boost the musical support.

All went well to begin with. The engine tooted its start signal, and eventually everyone arrived at the 'half-time' rest point in Sandbach. Overcome by the magnitude of the occasion, the landlord of the local hostelry offered the perspiring bandsmen free ale for their efforts. The Wheelock Temperance bandsmen considered this a vile insult to their beliefs and went home in a huff. The Sandbach town men took advantage of the thinning of their ranks, to such effort that within a short time the entire bunch was so inebriated as to be quite unable to distinguish a crotchet from a quaver. Consequently the second half of the proceedings, in which the procession returned to Elworth, was conducted with no music of any sort and inevitably fell rather flat, despite the frenzied efforts of the marchers with their uniforms and colourful floats to liven things up.

Such undisciplined behaviour appalled the gentry of Elworth, and their anger at having their fine plans spoiled by a few jugs of ale was scarcely tempered by the knowledge that the culprits were not solid Elworth inhabitants, but those rather inferior people from Sandbach up the road. That very same evening, at the Commercial Hotel in Elworth, a meeting decided that Elworth should have its own band so that they need no longer rely on outsiders for their musical support. Among those present at that meeting was Edwin Foden, and before the evening was out they had started a subscription fund to set up and equip a band. The day ended with the effigies used in the procession being thrown on to a hugh bonfire, and so high was the resentment among the Elworth folk that many were heard to cry that it ought to be the Sandbach band on the fire instead of the effigies!

Before the summer was out sufficient funds had been collected locally to buy a set of instruments, and such was the excitement when they arrived at the railway station that the bandsmen seized their cornets, brasses and horns and set off on a tour of Elworth, blasting away on their new instruments with lots of enthusiasm but precious little skill at that stage. Among them were Edwin Foden's two sons, William and E.R., and they were among the first to win their 'proficiency badge' in the shape of a straw hat that was given to each man when he became sufficiently skilled to play in public: in fact the hat was the only uniform issued.

The band was known as the Elworth Silver Band and showed a great deal of promise. They gave performances at several towns and villages in the area, and unfortunately learned some commercialism in the process. For when; in June 1902, the coronation of King Edward VII was celebrated in Elworth, the band demanded an

*Forerunner of the Foden Motor Works Band was the Elworth Silver Band, formed in 1900. This picture was taken just before the coronation of King Edward VII in 1902. Edwin is wearing a black hat on the right, with his son, William, third from the right on the front row. E. R. Foden with his cornet is second left on the back row. The three-ton wagon is decorated for the occasion, and is similar to the one which took part in the War Office Trials six months earlier.*

extortionate sum for appearing, thereby offending the town dignitarties who had created the band in the first place. In the event, the music for the celebrations was provided by a band from Crewe, and the Elworth Silver Band was shortly afterwards dissolved by its founding committee.

Edwin Foden was furious. Indiscipline from Sandbach folk he resented but put up with grudgingly. From his own people it was intolerable, and he declared his intent to form a new band under his own direction within the now-sizeable company of which he was the head. The company by that time was known as Fodens Ltd, and had great resources, not least a workforce of several hundred. He summoned the bandsmen, gave them all a good talking to and told them they could join his new band, but it was going to be the best in the world and, if they didn't like hard work and discipline, then now was the time to say so. A new set of Besson instruments was bought, smart uniforms replaced the old straw hats, and the Foden Motor Works Band was born.

As well as providing the 'Sunday-afternoon-in-the-park' type of entertainment that was so popular in Britian for the first half of this century, the new band entered brass-band contests, many of which took place in the north during the prewar period. For a long time most of the original musicians stayed on and they had a lot of fun, did a lot of travelling, and gained a lot of experience; but they did not win any prizes. This was scarcely surprising, when you consider that such already-famous bands as Black Dyke Mills, Besses o' th' Barn and Cresswell Colliery, to name but a few, were their opponents, as well as lesser but very experienced bands.

Their first success was in a contest coinciding with a flower show in a small Lancashire town, Haslingden, two years after the formation of the band. Edwin was delighted. His men were at last performing as he had wished, and this made it all worth while. These early band contests were based on marching ability as well as musical skill, and this brought its problems to non-military men like the artisans from Elworth. There is a tale from the band's early days concerning yet another competition failure, caused by an unfortunate drummer called Wright who, being small of stature and unable to see over the top of his drum, marched straight ahead beating enthusiastically while his comrades turned a corner, and became aware of his mistake only when the resounding brass sounds began to fade. But at that time, it was all taken as good fun, and the many disappointments were drowned in a glass of ale before the musicians returned home.

At about that time, the well-known conductor and composer of brass music, William Rimmer, heard the band play. He made known his opinion that the band was almost totally wrong in its make-up of players and needed a major re-arrangement. Edwin would not hear of this, particularly as his sons were leading players, but, when another prominent bandmaster, Alfred Jackson, agreed with Rimmer, the realisation dawned on the Fodens that Rimmer was right. Soon the necessary re-arrangement was put in motion.

Edwin Foden, by then 65 years old, was a perfectionist, and he decided that his band would have the finest players in the land. One of the main offices was turned into a temporary audition room, and advertisements were published to attract the best in the land. Applicants came in large numbers, some genuinely brilliant, others merely hopeful. Of the players who had been in the original band, most lost their places: among them were Edwin's two sons, William and E.R., his son-in-law S.P. Twemlow, and Fred Mason's brother, George. Commercial or industrial abilities clearly did not always coincide with musical skills. Having seen the extent of the reshaping of the band, William Rimmer consented to becoming its musical director and bandmaster, and under his extraordinary leadership it was but three short years before the Foden Motor Works Band won the British Open Championship. That same year, 1909, they were second in the British National Championships at Crystal Palace, London. Edwin Foden was well pleased with this, but was heard to say in private to Rimmer, 'Come and tell me when you've won 'em both.' Rimmer was, of course, no stranger to Edwin's humour, and he accepted the challenge as it was meant; as a compliment to his work, and as an implied vote of confidence in his ability to lead the band to winning the coveted 'double'.

Edwin Foden did not have long to wait. The

very next year, in 1910, the Foden Motor Works Band won first the Open, then the National Championship. Now that really was cause for a celebration. Half the county of Cheshire, it seemed, turned out to greet the triumphant bandsmen, just as they had when Edwin and William chuffed home in their victorious steam wagon after the War Office Trials nine years previously. So thick were the crowds that the bandsmen had difficulty in making their way back to Sandbach and Elworth, where a special party was to be given, and at each place along the way they were obliged to play a small part of the pieces that had won them the prize at Crystal Palace. The party arrived at Elworth more than an hour behind schedule, but nobody seemed to mind.

This time Edwin Foden had no reservations about showing his pleasure. In 1902 he had declared that he wanted his band to be the world's best, and now they were. To signify his approval he made a gift of five shillings to every one of the old people of Elworth, and in 1910 five shillings was a handsome gift indeed. William Rimmer had retired from the leadership just before the great triumph, and it was under the new baton of Edward Wormald that the success had been won. Few bandmasters can have had such a spectacular introduction to their work as Wormald at Fodens in 1910.

Sadly, Edwin Foden was to see no further success with his band, or for that matter with his wagons for, on August 31 1911, he died, 70 years of age and a happy man. To his employees, and the people of Elworth and the surrounding villages, Edwin Foden was a special and much-loved man. He had brought prosperity, comfort, security, community integrity, and entertainment, not to mention considerable national repute to their little corner of Cheshire. Even 50 years later, it was common to hear local folk talk of the days of 'Mr Edwin' with a reverence and nostalgia of rare intensity. It was against the background of this considerable respect and love that the band carried out one of its saddest tasks, leading the funeral cortège with a selection of Edwin's favourite Methodist hymns.

Despite being rejected as players when the band was re-formed in 1906, William and E.R. both retained a close interest in its running, and in Edwin's last days they took over as band manager and secretary respectively. These were fairly time-consuming jobs: the requests for concert engagements grew to such a number that the majority had to be turned down, and minor band contests were discontinued as well. However, a number of brass quartets were formed from among the band, and these proved to be both popular with the public and adventurous in their music.

In 1912 the band won the British Open Championship yet again, but the biggest honour as far as the men were concerned came in 1913, when King George V and Queen Mary visited the Marchioness of Crewe, who lived not far from Elworth. Their Majesties were very fond of music, and asked that the Foden band entertain them during the evening. Such was the pleasure of the King that he drew the somewhat overwhelmed Mr Wormald aside afterwards, and asked him to return with his men the next day to play for the royal party as they prepared for their departure back to London. Such royal approval for the band would have pleased Edwin. It also enthused the bandsmen, who promptly went on to win the Open for the fourth time.

Demands for the band's services were such that their travel arrangements became almost a full time occupation so, instead of hiring transport for trips, it was decided by the board of directors that the band should have its own transport. This was a hybrid steamer, comprising a three-ton chassis, on to which the higher-powered engine and boiler of the five-tonner model had been grafted. A local firm from Edwin Foden's home village of Smallwood built the body, which had a stylish cab, streamlined more for effect than efficiency, and an enclosed body terminating in a sort of gallery at the back, rather like an old-fashioned American railway carriage. The name *Puffing Billy* was on a plate over the smokebox, as was the practice with the names of most Foden steam wagons by that time.

On the vehicle's first day — to an engagement at Knutsford on May 1 1914 — E.R. Foden, who was first of all an engineer, and second a bandsman, travelled with the machine to make detailed studies of fuel and water consumption against speed, as he had the notion that a hybrid like this might have a useful sale as fast transport

## The Foden Motor Works Band

*The famous band bus,* Puffing Billy, *takes on water from a roadside standpipe. This machine was used as a mobile experimental chassis as well as transport for the band.*

for relatively light cargo or for passengers. His plans were ruined by one of the party who somehow lost his flugelhorn overboard while *Puffing Billy* was in full cry at something in excess of 18 mph. The resulting stop, and the long reverse to retrieve the instrument from the ditch, totally ruined E.R.'s test figures, and with it the pleasure of at least one bandsman's day out.

*Puffing Billy* was hired out to all kinds of organisations between band engagements, including local school outings, and for more than 25 years it was a familiar sight, chuffing gently along the lanes of Cheshire and frequently further afield into neighbouring counties. Finally the gallant steamer was broken up for scrap, and the band had to make do with hired transport again for a year or so until the new coach was built.

The year when *Puffing Billy* was built brought deeper problems, in the shape of the Great War and, whereas the bandsmen had previously been employed purely as musicians, they now had to join their fellows on the factory floor, turning out munitions and a variety of wagons and trailers for the war effort. This meant they had to rehearse after working hours, despite the fact that most men were doing 10-11 hours in the factory each day. Clearly under those conditions — which were suffered by all the other bands in the country as well — musical standards were bound to deteriorate, and for that reason the National Championships at Crystal Palace were suspended for the duration of the war. However, the organisers of the Open at Manchester decided that music was something everybody needed just as much in wartime, if not more so, as in times of peace, and they continued with their contest despite the difficulties. The Foden bandsmen managed to find time for a number of concerts, mainly at functions aimed at boosting the war effort, and with that stimulus they won the Open yet again in 1915.

But the success hid all manner of resentments, arguments and dissensions. A faction existed which thought that the band should have special treatment in view of the fact that they aided the war effort with their concerts as well as toiling in the factory. Such treatment was not forthcoming, however. In the dismal winter of 1915, when morale was at a low ebb, with huge casualty lists coming daily from France, the seemingly adverse trend of the war, and food and supply difficulties on the home front, a dozen bandsmen went on strike for special conditions, forming a picket line in front of the factory. Despite a great deal of sympathy with the difficulties of their employees generally, the Foden brothers and their fellow directors would accept no such action from any individual group. They were all in the same boat together, and everybody would have the same treatment, like it or not. Consequently the 12 were sacked on the spot, despite the serious loss of skills in the band.

By the end of the war the band was but a shadow of its former self, a state that applied to a great many organisations in postwar Britain. It was now under the direction of Thomas Hynes, who had the unenviable task of restoring it to the standards of 1913. The task was daunting, but Hynes stuck at it, and painstakingly rebuilt his band, until six years later it was approaching the standards of the prewar years. He was then well past retirement age, and it is no detraction from Hynes' achievement that the man to whom he passed the baton was to take the Foden Motor Works Band to its greatest ever successes. That man was Fred Mortimer, father of equally

*The band poses with its impressive trophies, including the Crystal Palace Trophy for the 'National' Championship and the Belle Vue Challenge Cup for the 'Open'. This picture was taken shortly before William Foden left for Australia in 1924. He is at the extreme right, with his brother at the opposite end of the row.*

famous sons Alex, Rex and Harry Mortimer, all known to millions, through broadcasting and live concerts, as arch-exponents of brass-band music.

Quite apart from his sensitive and inventive interpretation of other composers' music, Fred Mortimer was himself an accomplished composer, conductor and player. Perhaps his most popular public performances were those where he conducted the band in one of his own attractive arrangements of a well-known piece, wielding the baton with his left hand while playing a brilliant cornet with his right. With such a man in command, backed by the team which Thomas Hynes had so carefully assembled, success was more or less guaranteed.

But if this was a vintage period for music, it was very much the opposite in industry. The hope of the postwar years had faded and Britain was fast heading for its worst depression in modern history. Work was scarce, unemployment widespread, orders were difficult to come by, and even those who had jobs were on bare minimum wages, often working one week and then being laid off for a week or more in between. The bandsmen were no exception, and even some of the best players were obliged to play for pennies in the streets to help feed themselves and their families. But in some strange way, the privation, the uncertainty and the gloom of those depression days seemed to lift the musicians to even greater heights, perhaps because this was the one thing left in which they could put their trust.

By this time, William Foden and his family had emigrated to Australia, leaving E.R. as the only Foden in Elworth to carry on the family tradition. Despite the difficulties surrounding their trading position, E.R. saw that the band was properly looked after. Neither 1924 nor '25 brought major successes, even under Fred Mortimer's magic touch, but in 1926 the Open once again went to Foden. Repeat performances in 1927 and '28 made them outright winners of the Belle Vue Challenge Trophy, an achievement unique since the competition began in 1889.

Fred Mortimer took over the secretarial and managerial tasks, leaving E.R. free to devote all his sorely needed energies to the company's affairs, while Harry Mortimer became bandmaster. But this did not mean the end of E.R.'s interest. When the band won the British National Championship for the second time in 1930, E.R. was as pleased as his father had been in 1910. Remembering Edwin's gift of five shillings to each of the old folk in the town, E.R. repeated the gesture by giving each of them a ten-shilling note promising that this would become £1 when the band next won the National. Now ten shillings was a small fortune in the dark days of the depression and, although there were some who suggested that the money might have been directed to more needy causes, there was no denying the general appreciation of the gesture.

The £1 gratuity was not long in coming. Just two years later, in 1932, Fred and Harry Mortimer took the band to a third National Championship, and the Elworth folk once again enjoyed the victory in what was becoming the traditional manner. It was perhaps fortunate that E.R. made no promise on that occasion to double up the gratuity next time, because the band went on to win the National in 1933 and 1934, making three in a row. No record exists of exactly how many people received E.R.'s pound notes, but there must have been over a hundred; on that basis the 1934 payout would have exceeded £400 had it been made, and even E.R. would have been hard-pressed to raise such a sum.

Although, at the end of 1932, E.R. resigned, he still retained his interest in the band, and his £1 gratuity was in fact made as a retired citizen, not as a company official.

A curious anomaly arose soon afterwards. Together with his son Dennis, E.R. set up a new company in Sandbach in the spring of 1933 to manufacture the ERF lorry. Even after E.R. had launched his new company, which was effectively a rival to Fodens Ltd, he still actively supported the band. Indeed in 1936, when yet another National Championship was won, he showed his pleasure by a further gift to the old folk of Elworth, this time to the tune of £2 a head. Few people can have been managing director of one company, and simultaneously patron of the band of a rival company, but such was the character of Edwin Richard Foden that he not only managed to perform that delicate trick, but also retained the esteem of both sides in the process. To show that his company bore no personal grudge, E.R. put up a big flagpole outside his little factory in Elworth Road, Sandbach, and from it flew what was generally supposed to be the largest Union Jack in Cheshire to welcome the Foden Motor Works Band home after their National win at Crystal Palace in 1936. This was the last time that Crystal Palace was host to the Foden Band, because it was destroyed by fire that winter.

By the early '30s, of course, Fodens were well established in the manufacture of motor vehicles, and the ageing *Puffing Billy* had neither the performance nor the image to match the success of the band. Consequently a new means of transport was introduced in the shape of Foden's new diesel-coach chassis. Fodens built the body and, with a 4LW Gardner engine, the new coach entered service, albeit somewhat ignominiously. The coach's first trip was to Glasgow where it was to support the Foden exhibit at the Kelvin Hall show. Unfortunately it collided with a Glasgow tram on the final stages of the journey, but the coach drove on to its destination while the tram had to be removed by a breakdown truck — testimony to the ruggedness of those early Foden motor vehicles.

Another story about the new coach, called *Bandmaster,* arose from the manner in which it was lettered. The name was elaborately done in gold leaf and coach lining on the back of the vehicle, and being a modern design a large luggage locker was incorporated under the floor behind the rear axle. This led to the appearance of the name across the door to that locker, and in time this led to whispered jokes to the effect that this was where the bandmaster lived when the band was on tour. Being a bluff Yorkshireman, Fred Mortimer took all this ribbing with good humour, but it was not long before the offending name was removed to a less conspicuous place.

Honours continued to accumulate; in 1937 the Foden Motor Works Band became National Champions for the seventh time, and again the following year, so becoming 'three-in-a-row' champions for the second time that decade. A second royal command came in that year too, when they played at Windsor on St George's

Day. King George VI chatted to the bandsmen afterwards, while others explained the intricacies of the shining brass instruments to the young Princess Elizabeth, later to become Queen.

In the meantime, band concerts had become a very popular aspect of radio broadcasting, and the British Broadcasting Corporation had signed a long series of contracts with Fodens to play at regular intervals for millions of listeners. The band had appeared in films as early as 1935, and there must have been scarcely a person in the land who was not familiar with the name of Foden.

In addition to the National, the band continued to support other contests and events, especially in the north-west. Overseas tours came and went, and life seemed almost one long string of social and musical engagements. The company itself was prosperous once more in a booming society, with healthy export markets. Yet the dark cloud of war was approaching rapidly.

William Foden had returned from Australia in 1935 following a series of letters from the executives asking him to help reorganise the company. He remembered well the privations of World War 1, and was prepared for the second when it came. 'I will not', he wrote to E.R., 'permit the antics of some jumped-up German corporal to affect the running of my factory or my band.' That factory activity should be affected was inevitable, as long hours were spent turning out tanks, munitions and trucks, but the band played on. Indeed, the existence of such a skilled musical ensemble was seen by the War Office as a valuable asset in recruitment drives and in entertaining depressed troops and civilians. The savage nature of the war was brought home forcibly to the band in the summer of 1940, when they were giving a concert on the Sussex sea front. While they played, a straggling line of small boats came inshore, and tired, dirty, exhausted men struggled up the beach. It was the beginning of the Dunkirk evacuation of the BEF from France, and this sight moved many of the bandsmen to give up music and instead devote all their time to war production.

The war ground on for five more years, but as military success finally came to the Allies, the band made a tour of the devastated countries of western Europe, following close behind the advancing forces in the early months of 1945.

*The coach called* Bandmaster, *which replaced* Puffing Billy, *was hired out frequently. Here it sets off with a local Sunday-school party.*

## The Foden Motor Works Band

They took their now rather battered coach across on a landing craft to the 'Mulberry' harbour in Normandy, then played their way across France, Belgium and Holland, entertaining in hospitals, refugee camps, city squares and liberated prison camps. They arrived in Brussels on May 8, VE-day, and their concert that night in the Opera House was one of the best in their long history. The concert tour continued and finally ended with a gala concert in Paris, to the curious but delighted French, to whom brass-band music was not a familiar sound.

There was a great deal of work to be done after hostilities ceased, and William Foden, despite being 76, set about directing the necessary arrangements with considerable energy, and the skill born of long experience. He became governing director of the company, and of course set about re-establishing the band as a major musical force. The National Championships, suspended during the war years, were revived, with the Albert Hall in London as their venue. But throughout Britain the brass-band movement was ailing. Too many fine musicians had failed to return from the war, and others were insufficiently fit to take part in active playing. Furthermore, the higher standards of wireless broadcasting and the introduction of television to a mass audience led to a waning public interest, and people were less inclined to go out to a concert now that there was good entertainment at home at the turn of a switch.

Fred and Harry Mortimer were nevertheless intent on seeing their band at the top again, and they had the backing of the company. Occasionally advertisements would appear in the press to the effect, 'Wanted, radial drill operator, must be able to play cornet to high standard', and everyone knew that Fodens were working on both their engineering and their music. The reconstruction began to pay dividends, and 1950 was a good year both for the trucks, which were finding increasing success in new export markets, and for the band, which won the National Championship for the ninth time. Amid the triumph of yet another championship win, E.R. Foden died, and the band led the funeral cortège of the man who had played a major part in both the great truck-making firms in the Sandbach area.

The following year was 'Festival of Britain'

*In 1953 'Mr Willy' (centre) was still taking an active interest in the band. Harry Mortimer is on Mr Foden's left, with brother Rex next to the end of the row on his left. Their father, the famous Fred Mortimer, is on Mr Foden's right.*

year, and the band was busy the length and breadth of the country, giving concerts and supporting formal functions. They also won a special contest that year at Bournemouth which made them Festival of Britain Champions. In 1953, coronation year, the band was swamped with requests to support all manner of celebrations. Inevitably some had to be refused, but many were undertaken, including a programme of concerts in London parks in which they were joined by other famous bands like Fairey Aviation and Black Dyke Mills. It was during this series that Fred Mortimer died at his home in Elworth, having been too ill to join the band for the events in London. Undaunted, the band went on with Harry Mortimer to their tenth National Championship a few weeks later. Harry had established good contacts throughout the entertainment world, and as a result the band found itself involved in more and more radio work and an occasional television performance. This work did a great deal to revive the popularity of brass-band music, and for a number of years there was always a considerable public demand. There were even a few new bands on the scene during that time, one of which was established in 1955 by Fodens' competitors in the truck business, Leyland Motors Ltd, from Lancashire. One of Harry Mortimer's ideas was the 'Men o' Brass' massed bands programmes for the BBC, in which he combined the Foden, Morris Motors, and Fairey Aviation bands with very impressive results. Another of his ideas was to have dance-band leader Ted Heath conduct the Foden band in the *This is Your Life* television programme. That was in 1958, when the band were National Champions for the 11th time.

By this time the old coach from the early '30s was outdated and a new one, powered by the supercharged two-stroke Foden engine, took its place. (She is still in service at the end of the '70s, and shows little signs of tiring.) William Foden, although well into his 80s, still took a great interest in the band with which he had once played, and was invariably there with his congratulations when the band won yet more contest honours. There were tours at home and overseas, and more and more broadcasts. Performances ranged from one extreme to the other. In 1971 they gave a superb concert in York Minster, as part of the celebration of the 1,900th anniversary of the founding of that great city, and in the same year they appeared with 'pop' musicians Georgie Fame and Alan Price in a special show at the famous London night club Annabel's.

In 1974 Harry Mortimer finally decided he was too old to continue as active musical advisor, and his brother Rex retired as musical director the following year, so ending more than 50 years collaboration of Mortimers with Fodens. The band somehow had never seemed quite the same since the death of William Foden in 1964 but they managed to win the Open in 1964 and came third in 1970. With the retirement of the Mortimers, it seemed as if there was no motivating force left and, when John Golland took over from Rex Mortimer, his was an unenviable task. The severe commercial crisis experienced by the company was a further blow, and though the band was wisely retained it was nothing like its former self. The full-time professional musicians and an elaborate and expensive *modus vivendi* both had to go.

But the company survived the crisis and grew once again to stability and strength. The band did likewise, and is once again entering contests, particularly the Open at Belle Vue. The Queen's Jubilee year of 1977 was a busy one for the band, with performances at the Albert Hall among many other venues. Sandbach School reached its third centenary at the same time as the Jubilee, and the band played at the school's celebrations.

1978 saw a return to competition success as the band won the BBC 'Champion Brass' contest in June.

Few companies can claim such a long history of musical as well as technological success as Fodens, certainly not in the vehicle industries. Many see the brass band as a dying breed, but at least two companies disagree. One obviously is Fodens, the other is British Leyland, which in 1978 revived its brass band in order to make good music and to help promote the company in fields outside engineering and commerce: a splendid idea of course, but scarcely original. Edwin Foden had just such an idea in 1902.

# Chapter 12

# The spawning of ERF

Towards the end of the 1920s, things had not gone well for Fodens Ltd, and indeed it seemed that luck had deserted the company when, in 1924, Mr William Foden, oldest son of the founder, severed his connections with it after being commercial and joint managing director for many years, and went to Australia with his family to take up farming. His younger brother Edwin Richard, known to everyone simply as E.R. (the name Edwin was used exclusively for the company's founder), remained at the works as a director in charge of sales, with his brother-in-law Mr S. P. Twemlow as the other remaining director from the days when 'Mr Willy' was there with them. E.R.'s son Dennis was a young salesman with the company too, but he was far too junior to have a board seat.

The popularity of the traditional overtype steam wagon with its engine over the boiler was waning, indeed it could be said that the steam wagon itself was nearing the end of its long and colourful existence, faced with real competition from the internal-combustion-engined lorry. Although Fodens continued to build steam wagons in diminishing quantities until 1934, and the Sentinel Company in Shrewsbury continued even later than that, there is no doubt that, even in the mid-20's the steamers' days were numbered.

Demands for more loadspace better weight distribution, a better driver's cab, and higher performance, led to Fodens' producing a series of 'undertype' steam wagons. Although most of them worked tolerably well — especially the Speed-Six model — they were never as successful nor as reliable as the allegedly outdated overtype with its locomotive boiler had been.

The depression which hit all of Europe and America in the late '20s was at its worst, orders for vehicles of any type were extremely scarce, and morale was low as week by week more and more employees were either laid off or dispensed with altogether. An illustration of the efforts made by E.R. to cause as little hardship as possible can be seen in a work-sharing scheme he attempted to organise in 1928, whereby employees would work one week in two, so that as many as possible could have some kind of a wage and avoid drawing the 'dole', which was pitifully small in those days.

At the beginning of the new decade the company chairman was still Francis Poole, who had held the position for many years, and Mr Wood Whittle was general manager. He was the latest in a succession, none of whom was particularly successful in pulling the company together. Determined to fire up some enthusiasm among the transport users, Wood Whittle ordered the building of several unorthodox new steam wagons, all with undertype layout, and one with an all-welded boiler. He also dabbled in internal-combustion-engined chassis, more as a research exercise than as a serious attempt at switching to this means of propulsion. Eventually a successful oil-engine chassis did emerge, largely at E.R.'s instigation, powered by a vehicle-conversion of the Gardner L2-type marine engine, after several disastrous adventures with Austin and Dorman engines. These failures and partial successes stirred up considerable bad feeling within the firm, as is invariably the case in such circumstances. It is merely human to go along with the crowd when things are going well, but indulge in recrimination when they do not. The staff at Elworth were only human; so were the management.

The Salter Report on transport only made matters worse. In the midst of all these problems it dealt a severe blow to the already shaky heavy steam wagon by imposing crippling taxation on any vehicle over four tons. The motives behind this report, which had government backing, and the identity of those responsible for its recommendations have always been a subject of bitter debate, and they are discussed in more detail in Chapter 13. It was clear from the report that there was little use for a steam wagon of the type that Fodens could build, except in special applications for some export markets. Amidst this despondency, the early Foden oil-engine trucks were born, after a great deal of trial and error on the part of the many good engineers still left at Elworth. Unhappily, the 'error' part of this process met with stern disapproval at board level, despite the fact that no guidelines existed as to how this type of truck should be built. A steady stream of perfectly competent engineers, seeking solutions to totally new problems, found themselves dismissed from their jobs — a disastrous situation for them, as new jobs were impossible to find. This practice was totally abhorrent to E.R. He had been with the company all his working life, so many of the men involved were personal friends of his, and he felt very bitter about the dismissals. However, his protests to the board met with little sympathy. E.R. soldiered on for many months. He worked with Ernest Sherratt, a promising young designer who had spent a number of years on the factory floor building both steam and internal-combustion vehicles. Gradually they overcame most of the technical problems and evolved a successful diesel or oil-engined truck chassis. But the personal conflicts remained, even worsened, and towards the end of 1932 E.R. finally resigned. He was 62 years old anyway, ready for retirement, and

*The design of the first lorry originated by E. R. Foden and his son Dennis was remarkably advanced for 1933, and the finished machine was extremely effective. Commercially it was an instant success, and set the new company on a direct route to profitability.*

*The spawning of ERF*

**Right** *Among the pioneering work carried out by the new company was the development of the heavy articulated tractive unit.*

**Below right** *Significant wording in the early public announcements of ERF activity sought to emphasise the total break with the old firm and its methods.*

wanted no further part in the bitter wranglings that had become commonplace at Elworth.

His son Dennis, on the other hand, could not afford to resign, but by the winter of 1932-33 Dennis had made up his mind that he could not let things ride. The situation at the works had deteriorated so much that, despite the huge unemployment total, men were resigning voluntarily, although a trickle of oil-engined trucks was emerging at last. Dennis travelled to Blackpool, where his father lived in retirement, and they talked about their ideas for the future. Their vision was of a range of very modern oil-engined lorries of unprecedented efficiency, with stylish and comfortable cabs, running on crude oil, which at that time cost 2¾d a gallon, or a fraction over 1p in decimal currency.

The idea germinated rapidly. Why not go ahead and do it, they asked themselves? Why not indeed? E.R. had a reasonable amount of capital, his two daughters were prepared to contribute, and young Dennis, who was probably the most enthusiastic of all, was more than willing to chip in his small contribution. They talked to Ernest Sherratt, who had the right blend of practical and design experience with Fodens to enable him to undertake the design and engineering activity, and he was willing to join them. They also talked to George Faulkner, another Foden man who was related to Dennis and E.R. by marriage. He had also resigned and would make an ideal works manager.

At Easter 1933 they had their first formal

## WHAT IS BEHIND THE E.R.F.

MR. E. R. FODEN'S first connection with Transport was in the year 1878 (when only eight years old), driving a traction-engine belonging to his father, the late Mr. Edwin Foden. In 1881, the first traction-engine built in Sandbach had many of Mr. E. R. Foden's ideas incorporated in it. He designed the first steam-wagon (on steel-tyred wheels) in 1898, and vehicles of this type had a very successful run until 1913, when solid rubber tyres came into favour. Subsequently he introduced the first pneumatic-tyred steamer, but as steam transport appeared to be going out of favour about 1931, Mr. Foden turned his activities to the production of a 6-8 ton chassis fitted with the famous Gardner oil-engine.

Owing to unforeseen circumstances arising in July, 1932, he decided to take a well-earned rest. After being in retirement for some few months, however, many requests were received from his old customers and friends to go into business again. Meanwhile, legislation had decreed that the maximum load-carrying vehicle to sell in any quantity must have an unladen weight of not more than 4 tons. Mr. Foden and his son, Mr. Dennis Foden, therefore decided to manufacture vehicles of this taxation class and embodying the strongest frame, powerful and reliable engine, robust axles and gear-box, and efficient brakes.

These vehicles have been built on totally different business lines, thereby reducing production costs very considerably, and in that way our Customers are reaping the benefit.

It will give us the greatest pleasure and be much in your interest to let us quote you the next time you are considering the purchase of commercial vehicles.

meeting, held in a lean-to greenhouse at the side of a house belonging to E.R.'s daughter, opposite the Foden plant. They decided to go ahead and produce their own oil-engined lorry, and a new company was formed — E.R. Foden & Son. The capital of the company was made up from £12,000 contributed by E.R., and a further £8,000 split between Dennis and E.R.'s two daughters. At this stage it was not a limited company.

The difference in the men's role is interesting. Although E.R. was the senior man with most experience and skill, he chose to be a 'figurehead', lending his great authority and high standing in the industry to the new venture, so that suppliers and potential customers would grant it full credibility. It was young Dennis, just 33 years old, who was the driving force behind the whole new concern. It was he who toured the north-west and the Midlands of England in his Alvis motor car, talking to such people as Tom and Joe Gardner at Gardner Engines in the Patricroft district of Manchester, Fred Cowell, the commercial director at the Kirkstall Forge Company (makers of heavy duty axles), Frank Brown, the transmission manager at David Brown Gears, and many others. Such was Dennis' enthusiasm, fired no doubt by his relief at being out of Fodens, that he talked all these suppliers into providing the first few sets of components in a provisional order of 25 sets on a 'sale or return' basis. This remarkable concession reflects the reputation that E.R. had acquired over the years.

The beginnings of the new lorry were modest indeed. Ernest Sherratt had a drawing board made by a pattern-maker friend, on which he drew the layouts and details of his new lorry. The working prints were made from these drawings by exposing them over the sensitive ferro-prussic paper in an old borrowed picture frame. Slowly the design took shape. Dennis' wife Madge Foden did all the letter writing, paperwork and bookkeeping, sometimes working in a small corner of the greenhouse not occupied by drawing boards or printing gear, but more often in the lounge of their home.

Meanwhile, George Faulkner, whose job it was to actually build the new lorry, toured the district recruiting the best men he knew, many of whom had either resigned or had been dismissed from the big works across the road. He then had to find somewhere to build the chassis, because the greenhouse was clearly nowhere near large enough for that purpose, although it provided adequate if somewhat warm and steamy office accommodation as the summer wore on. As luck would have it, the coachbuilding firm of J.H. Jennings in Sandbach was given notice by a firm of baby-linen manufacturers that rented a portion of the works, that they were moving to larger accommodation. George Faulkner went to have a look at the place, and found it spacious, dry, and above all well lit from large windows in the roof. It was ideal for the new company's limited scale of operations and the workshop was rented on the spot. It is ironical that Jennings was eventually taken over by the same company that moved in on one exciting day in June 1933 with its ten new employees; but that is another story.

Work began on building the prototype chassis immediately. So well had Sherratt done his part that the very first lorry went together like a designer's dream, with no major snags and no delays. And so swift was the process that by September 1 the vehicle was finished, fitted with a dropside body, and tested, ready for its new owner. Madge Foden demonstrated just how easy it was to manage by starting up the brand-new machine, and driving it away.

The new lorry by E.R. Foden & Son was one of the star exhibits at the London Motor Show in November. By then three more vehicles had been built, and all were sold before the show opened its doors. Enough orders were taken during the show to establish the success of the venture beyond doubt, and two dealers were signed up: W. J. Boyes in London and J. Brown & Sons in Edinburgh. It is perhaps significant that both firms are still in existence today, and are still distributors for the Sandbach lorries.

The wording of the first advertisements contained a significant passage; ' . . . these vehicles have been built on totally different

*E. R. Foden was a keen motorist as well as a competent industrialist, and he enjoyed being photographed with his cars. His first car is difficult to identify, possibly an early Lanchester, but it was followed by a De Dion Bouton in about 1906 and then a Crossley in 1913.*

The spawning of ERF

*One of the latest and most successful ERF models — the B-series European tractor — is seen here at speed on a Belgian highway. Cummins engines giving up to 350 hp provide the motive power.*

business lines . . . ', emphasising the break with Fodens Ltd. Such was the success of the new make that the old firm asked E.R. to choose something other than E.R. Foden & Son for the title of his company and the name on the front of his vehicle. Not wishing to prolong old differences, E.R. agreed, and the truck became known as the ERF. Shortly afterwards the company name was changed to ERF Ltd. It continued to thrive under the energetic direction of Dennis Foden and the wisdom and advice of E.R. Neither man is alive today: E.R. died in 1950 and Dennis only ten years later, but the company they founded is still very strong and is the largest builder of heavy tractive units in Britain. The original truck, which was sold to Fred Gilbert (a haulier at Leighton Buzzard in Bedfordshire), is still in existence. It was recovered by ERF a few years ago and has been restored to its original condition. The truck had worked for 30 years or more, and is by no means the only ERF to have done so. It does not carry chassis-number one, as might be expected, but number 63. Why 63? That was E.R.'s age at the time of its creation, and it is typical of the self-effacing Dennis that he should suggest that his father's age be chosen for that first chassis number, even though it was he, Dennis, who was the main driving force behind the truck's creation. The production chassis numbers thereafter followed on from 63, not from 1 or 001 as did most other makes. At that time Ted Foden was selling ERFs, but he soon returned to Fodens when his father returned.

Commercially ERF was almost a model company. So successful was the original truck design, and so well was it developed and managed over the years, that the company made a profit throughout its production with the exception of one half-yearly result which showed a small loss in the middle of a major model change. This loss, however, was well and truly swallowed by that year's final result, and ERF built up to multi-million-pound profits by the late 1970s.

The full history of ERF is a long and colourful story, which cannot be told here. The author's book *ERF* deals with the subject, and is available from the publishers of this volume in their *World Trucks* series.

# Chapter 13

# A painful transition

After the departure of William Foden in November 1924, his younger brother E.R. Foden was left as the sole remaining upholder of the family tradition and, with Sam Twemlow to help him, he struggled on against mounting economic difficulties and a declining market. Plans were laid with some optimism for the very modern undertype wagon, which proved disappointing in its initial form, although it eventually developed into a thoroughly satisfactory machine. Since the company had been re-formed into Fodens Ltd in 1902, it had done nothing wrong in the eyes of the customers, and some had sufficient faith in the new E-type wagon to buy it straight from the shop floor, without waiting to see how good it really was. This was something of a mixed blessing because, although this touching faith enabled the company to show a reasonable trading picture in 1925, it rather backfired on them when the later and much better E-type proved difficult to sell in the critical 1926-28 period because of the problems experienced with the earlier machines

Be that as it may, the directors were able to tell their rather agitated shareholders at the 1925 Annual General Meeting that their dividend was up from 5 to 6½ per cent, although gross profits were barely holding their own at just over £30,000.

If 1925 worked out rather better than anyone had dared hope, 1926 immediately shattered any remaining dreams of a steady return to easy and profitable economic times. The year began with a row in Parliament over the future of road transport, centred around the misappropriation of money from the mounting Road Fund held by the Treasury. Mr Winston Churchill was Chancellor of the Exchequer at the time, and he seemed to be as anti-road as any of his predecessors had been. A deputation from the motoring and commercial vehicle users' organisations went to see Mr Churchill, pointing out that, in 1920, government assurances had been given that none of the money extracted by direct taxation of motor vehicles would be used for non-motor-industry purposes. In particular they asked that the Road Fund should not be used for general treasury purposes while there remained so much work to be done on roads, including the repair or strengthening of more than 700 bridges in different parts of the country; these were in such poor condition that many had to be closed to motor traffic or otherwise restricted. Mr Churchill's reply sounded encouraging, but was in fact a masterpiece of the veiled promise. He said, 'I would not in any way accept the constitutional theory that Parliament has not the right to deal as broadly as it sees fit with the interests of the country as a whole. Nor do I accept that the Motor Tax is a tax paid voluntarily, so that they shall have the right to say how this tax should be spent. It should be regarded as a lamentable and preposterous act of folly if we were to cripple and wound the whole development of this wonderful new means of transport which is one of the greatest features of our lifetime, and which unquestionably by its smooth and speedy intercommunication has added vastly to the unseen and internal trade of the country.'

But any hopes which that reply might have raised, were not to last long. The following month the railway companies launched a campaign aimed at appropriating all national and local authority expenditure on road works of any kind, and charging it against direct taxes on road vehicles — excluding those lorries operated by the railways themselves. Mr Churchill was clearly sympathetic, and in his budget that April he earmarked over £7 million from the Road Fund for general-purpose spending by the Treasury. It was the beginning of a long series of such raids into the Road Fund, which since that day has never been used for the purposes originally intended by Parliament. This was a blow to the motor-vehicle movement in general, Fodens included. They subscribed to the view that a policy of neglect of the highways would slowly push the heavy commercial-vehicle industry into a backwater and that only with progressive road policies could the commercial vehicle realise its potential contribution to the economy.

But, if the 1926 budget was a blow, there was worse to come. There had been growing unrest in many industries, notably in mining and it was widely thought that there was a great deal of sympathy with the miners from other workers. On May 1 the miners went on strike. The huge weight of support that they enjoyed at that time can be judged by the fact that only three days later a general strike began. One daily paper recorded the event thus: 'The General Strike begins today, and when it will end no man can say. But how it will end there can be no doubt. It will fail because the Government and the people are resolute that it shall not succeed. They will repel it as a blow aimed at the very vitals of the State, and at the very existence of Constitutional Government and of Public Freedom. Moreover there is yet another reason why it must not only fail but fail soon, and that is that the vast majority of those who will come out in obedience to the summonses of their Unions have no heart in the struggle.'

This comment in *The Daily Telegraph* proved remarkably prophetic, and within a week the General Strike had collapsed: but while it lasted it gave the nation a fright such as it had not had since wartime days.

The major problem during the strike was that of food and supplies, most of which were immobile in docks and railway freightyards. The commercial vehicle showed, perhaps for the first time publicly but certainly not for the last, just how effective it was for the movement of supplies. Convoys of lorries with armed soldiers aboard carried food from the docks and the idle rail warehouses, and distributed them to the shops, via great transit depots set up in London's royal parks. The lesson was brought home to the people that the lorry, no matter whether steam- or petrol-driven, was not a thundering menace to life and limb, as many still contended in the columns of the press, nor a giant oppressor manned by the captains of industry regardless of the safety and comfort of the common man, but a valuable and vital contributor to the standard of life to which everyone had become accustomed. The General Strike did not affect Fodens

**Right** *The N-type wagon was the unsuccessful steam 'swan-song' at Elworth and only one was built. Despite its advanced design and high performance, it was not accepted by a sophisticated transport industry. The fireman worked with his back to the windscreen.*

**Far right** *An N2 wagon was designed but never built. The main difference from the one and only N was that the boiler was right at the front, so the fireman faced the same way as the driver.*

Ltd directly, because each and every man stayed at work, despite a certain amount of sympathy with the unfortunate lot of his fellows, particularly the miners. The loyalty of the men was noted in the annual report that year. As we have seen, it was probably as much the fear of losing their jobs, badly paid though they were at that time, that kept the men at their benches, as any higher motive of loyalty to the company.

Despite the troubles on the industrial scene, the 1926 balance sheet showed an increased gross profit of £40,473, although this was the result of a great deal of wage-cutting, and of a long list of minor property deals, so that only a small fraction of this figure was a real trading profit. Sales were hard to come by, and the 'deferred payments' figure grew to well over £90,000, the highest it had ever been in the company's history. The depression was really taking hold, and times were very unhappy at Elworth. The despair of E.R. at his increasing impotence in the firm is illustrated by his reply to an old friend's son, who went to him for a job — any job — in 1927. 'No good asking me, lad — I'm not the boss anymore,' he said with uncharacteristic lack of sympathy. He was, after all, still managing director, but rarely did he see eye to eye with his fellow directors.

The depression dragged on into 1927 and 1928 with just a crumb of comfort for Fodens Ltd, for the tax on petrol was further increased, giving the steamer a small additional lease of life. At about that time, the annual accounts published by the company began to get very vague, and it is difficult to see how the shareholders could have been persuaded to accept them. Such items as 'work in hand and goodwill as estimated by the managing director' would never have been accepted in former days, and neither would the omission of net profit figures. Trading profits, if any, were very small indeed, and what profits there were arose mainly from what were called 'sundry investment dealings' — mainly the selling of company houses.

By 1929 the gross profit had shrunk to £12,224, only £1,945 of which came from actual trading, while sundry debtors rose to an unprecedented £114,000. By then the Speed-Six and Speed-Twelve undertype wagons were Fodens' main line of business, although a few overtypes continued to be made, and a lot of them were returned for overhaul and reconditioning, which at least provided work of a kind to keep the company ticking over.

The only real ally that E.R. had during those difficult years was his brother-in-law, Sam Twemlow, but in 1930 Sam died, leaving E.R. as the sole survivor of the original team. John

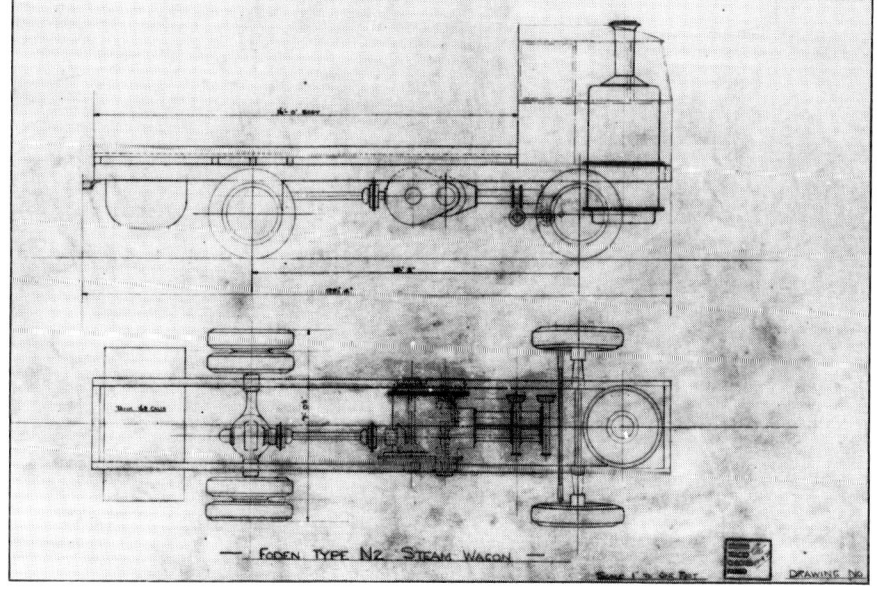

Stubbs, son of one of the original Stubbs brothers, came on the board, and for a time was a useful ally to E.R. The board had a whole gaggle of new men, including James Cowapp and Albert Jackson, who, although undoubtedly good commercial men, had little of that in-bred 'feel' either for the vehicle industry or for the men who made it 'tick'. Transport men, particularly truck men, are something of a special breed, and the board of Fodens Ltd in the early 1930s lacked both an insight into the characters of the workers and into the future path of the transport business.

Looking back with the advantages of hindsight, it is difficult to see how the board could have misread the signs that were there for all to see in 1930. The large industrial diesel engine had long been a reality, indeed industrial diesels had been installed at Elworth in 1924 to provide internal power and light, replacing the steam engines which had served since the beginning of the century. In Germany MAN and Daimler Benz had developed the small diesel engine to an acceptable level for use in a motor vehicle, while in Britain, Gardner, Dorman, Crossley and others had developed their own versions of the new 'oil engine' for industrial and marine purposes, and all manner of companies from small hauliers to big manufacturers were investigating the oil engine as a means of motive power. The diesel engine's main advantage was that of economy, since operating cost had been the major bugbear of the steam engine for many years. Not only was the steamer time-consuming in the extreme, but it made relatively inefficient use of the fuel it burned; no more than about nine per cent of the thermal energy contained by the fuel ever reached the road wheels in the form

*One of the more unusual machines was a rail car using the E-type engine. It was built at Elworth in 1927.*

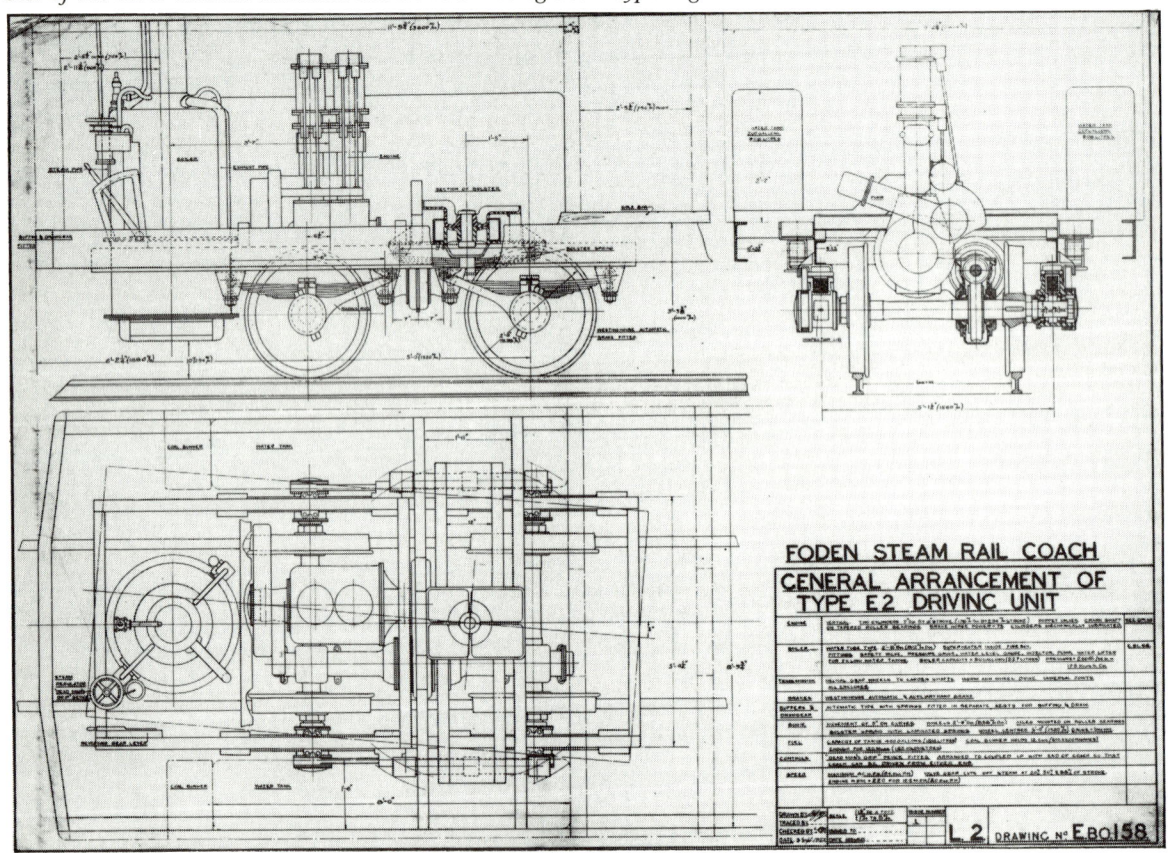

of tractive effort. The diesel, on the other hand, returned something like 33 per cent on its energy input, even in its infancy, and such a potential cost advantage was not lost on the far-thinking men in the transport world.

Among the men who were aware of the diesel's advantages was E.R. Foden. He had always been a keen motorist and, although the business of his company had always been the production of steam vehicles, he never lost sight of the fact that there were other means of making a heavy vehicle move, and believed that any new power source should be closely examined and evaluated. Unfortunately his views were not shared by his fellow directors, many of whom were extremely conservative — even reactionary — in their outlook, and they regarded E.R., who was by then over 60 years old, as at best a silly old fool, and at worst a heretic. Had not the company, after all, done very nicely thank you as practitioners in steam for sixty years? So why should it change now just because times were a little hard?

At the AGM the following policy statement was included in the chairman's speech: 'Now Ladies and Gentlemen, there is another matter which has been mentioned to me several times, and that is, why we don't go in for Oil Engines. In the first place we are all steam men at our Works, and we have now produced a steam wagon second to none, in fact it is in a class by itself and can hold its own against any Oil Engine made. (*This was the ill-fated Q type.*) It is cheaper to buy, cheaper to run and uses British fuel which must help British Industries. As you know, money staying in this Country must circulate and we have a chance of getting some of it back. On the other hand Oil Engines of all descriptions use a fuel which sends at least 75% of the money abroad and it is a thousand to one we shall get none of that money back.

'Another reason is that all the makers of the Petrol Engines and Crude Oil Engines have a great start on a firm which has never made a petrol lorry, and we should be years before we could compete against them. We should require new Engineers, a lot more money spent on new machines to compete against the present Oil Engine makers, and we should have at least twenty more competitors to face, therefore I honestly think we should be better keeping to steam for it has not yet got to its best. I really do think that we now have with our new Steamer, a wonderful chance of getting back business from the petrol people, as in all our tests against Petrol Lorries we have equalled them in speed, in fact with 10 Ton loads we have beaten them and as regards economy, we have saved between £5 and £8 per week, which is bound to tell in the long run. If trade generally will only improve we feel sure that we shall get our share and have a different tale to tell next year.'

But E.R. refused to be silenced and he finally managed, in 1930, to get the board to sanction a pilot programme of investigation into the diesel engine. The programme was tackled energetically by E.R. and fellow-engineers like Edward Mason and Ernest Sherratt, but there were difficulties. The main problem was that there were no guidelines to how the job should be done. The problem was an exact parallel to that which faced Edwin Foden and young William in 1899 when they set about building their first steam wagon. Some very peculiar devices had appeared at that stage, and it was reasonable to assume that some equally peculiar contraptions would emerge before they got it right this time. There was, for example, the choice of engine; should it be diesel-powered, or was petrol worth looking at as well? And which engines? There were several available, and various people spoke well of each one in some installation or other. Hard facts and figures were difficult to come by, and in such circumstances the only way to develop fact is to experiment. This unfortunately takes time, and time was not on E.R.'s side.

In August the board quite gleefully announced to its shareholders that they 'had investigated the matter of the diesel engine', and had found it 'inadequate and unsuitable' for the quality of commercial vehicle that Fodens Ltd was interested in building. E.R was furious, and even more furious when he saw the way the balance sheets had been prepared. There were no stock write-downs that year, thus making it look as if assets were worth more than they actually were; a big slice of funds had been written off against his diesel experiments, and the trading profit was only £8,000.

It required more than a temporary set-back

*Left The first diesel truck happened to be the most successful. It was built in 1931 with a 5L2 Gardner engine, Daimler gearbox and Kirkstall axle. Later it had a Foden box fitted and it was subsequently re-cabbed. The chassis number was 14004, and it was delivered on October 14 1931.*

*Below right A less successful motor truck was the two-tonner, powered by the Austin Cherub petrol engine. The main problem was engine and drive-line durability.*

like this to put E.R. off the scent, once he had made up his mind. Although he continued to work on development work with 'advanced' undertype steamers, most of which were essentially obsolete before they had been completed in the experimental shops, he clung to his belief that the diesel truck was the way to go. Just how he persuaded the board to sanction the work is not recorded, but many who were working at Elworth at the time think that much of the work was done simply on his own say-so, and was presented to the board as a *fait accompli.*

Small 30-cwt and two-ton chassis were tried as part of the trial programme, with either Austin 'Cherub' petrol engines, or Dorman 4DS, Dorman 4JUR, Meadows 4EL, Beardmore, and Russell & Newbury power units. Fortunately one of the first engines to be tried happened to be the best. In the nearby Patricroft district of Manchester the Barton Hall Engine Works, run by a family called Gardner, had developed a range of compact direct-injection diesel engines, following extensive experience with industrial gas and oil engines. The Gardner engine was called the L2, and was available with two, three, four, five or six cylinders, depending on how much power was needed, all cylinders and pistons being identical.

Development of the undertype steam chassis followed quite closely that of the heavier internal-combustion-engined chassis elsewhere; in fact the axles and frame of the Speed-Six and Speed-Twelve were entirely conventional by 'lorry' standards. Consequently the finding of a suitable chassis was not difficult, nor was fitting it with suitable axles and springs; all these parts were, in fact, Speed-Six components, some with minor modifications, but most of them perfectly standard. The Gardner engine was installed up front on solid mountings and, as in the Speed-Six, a short shaft ran back to the gearbox a little way along the chassis. The type of gearbox used in the steamers was not suitable for the diesel, so a Daimler bus gearbox was fitted to the prototype. A second short shaft ran rearwards to the axle, and the brake system on the rear wheels only, with four shoes in each drum, was identical to steamer practice. Steel wheels with 40 × 8 tyres completed the chassis, and an angular coachbuilt cab topped off the new creation.

The new lorry not only worked admirably, but it also caused a great amount of interest in the outside world. The board, however, was displeased — E.R. was retained on the board, but only after severe censure. Just to emphasise the point, a new advertising and promotion campaign for steam wagons was launched, and was remarkably effective in evaporating most of the interest which E.R. had succeeded in raising in

*A painful transition*

his new diesel machine. It should be remembered that in 1931 few manufacturers in Britain were offering diesels, even as an option. Leyland, Thornycroft, Albion, Dennis and the other UK heavy-lorry builders were still firmly wedded to petrol engines, and the few diesels around came mainly from Europe, although a few Gardner and Dorman conversions were being carried out here and there. Yet the economy of the new engine was astounding. There is little doubt that, had E.R. been permitted to develop his new lorry as he had wished, the Foden name would have recovered its position in the market very much sooner and, who knows, may today have been a much bigger power in the truck world. But this is mere conjecture: the fact remains that the Foden board did not want to know about diesels, and the Works and General Manager, who had been luckless enough to have witnessed E.R.'s work, was dismissed and replaced by another man, called Wood Whittle. His was a name that would eventually play a major role in affairs at Elworth.

Despite the attitude of his board, E.R. decided to soldier on, quite convinced that he was working along the right lines. Somehow he managed to build a small series of diesel trucks, all of which found ready customers. The first one, built in 1931, was bought by Jacksons, hauliers of Willaston, and worked there for many years, later being fitted with a Foden gearbox instead of the Daimler, and a new and more stylish cab: the vehicle has been preserved in this form. The big L2 Gardner was a little on the bulky and heavy side, according to E.R.'s wishes, and it was at this stage that he experimented with the various other engines already mentioned, mainly in a light chassis with about two tons carrying capacity. Unfortunately most of those efforts ended in disaster, partly because the components were not really suitable for the heavy work intended, and partly because the users thought that these little machines ought to be able to work like the old heavy steamers. The result was that this new motor vehicle, launched in a half-hearted manner because of opposition within the company, lost a great deal of money. That year of 1932 saw the first big loss in the company's history, with a trading loss of £48,351, and further losses on investments and property bringing the total up to £75,941. The shareholders were aghast, and there were many probing questions at the AGM in the Crewe Arms Hotel that August. But the results were inconclusive, and still the company muddled on, undecided whether to continue making steamers — which were built in extremely small quantities by that time — or whether to make the full transition to diesel vehicles, or indeed to petrol vehicles.

Within the works morale was at a very low ebb. The catastrophic losses had led to a demand from the shareholders for stringent economies, and much of that economy took the form of the dismissal of workers of all grades on the slightest pretext. There was a random selection of those to go, with little thought for preserving the jobs of key men so that some semblance of a team could be retained. E.R. did what he could to retain the best men, and ended up in virtual open warfare with Wood Whittle, the general manager, in whom the board had invested powers way beyond those normally carried by such a position. The final straw came once again in the form of a legislative blow.

In 1929 the government of Mr Stanley Baldwin set up a series of Royal Commissions to look into the whole question of road haulage. The first merely suggested that more control should be exercised in the interest of public safety without saying exactly why or how this was to be done. The second was more positive in that it proposed a system of carriers' licensing as a form of control for the industry, and as such was incorporated in the 1930 Road Traffic Act which formed the basis of transport licensing for nearly 40 years, with Traffic Commissioners controlling the whole system from regional offices. This did not go down too well in certain circles, although it must be said that these measures were needed to control some of the less responsible elements in the haulage industry. The third report of the Royal Commission suggested that severe weight restrictions be imposed on commercial vehicles by means of tax sanctions. This set the feathers flying, both in Parliament and in the transport industry itself, the main argument being that it was unfair to penalise heavier vehicles, since they shifted cargo more efficiently than smaller vehicles. So great was the outcry that the National Coalition Government under Mr Ramsay MacDonald, which followed Baldwin's administration in 1931, was forced to do something. That something was yet another committee, under the chairmanship of Sir Arthur Salter.

The official objective of the committee was to examine the relationship of road haulage with other sectors of industry and transport, and in particular to examine its financial obligations.

*Curious-looking machines were the E-type export tractors, built round the E-type wagon mechanical units. Most of them went to South Africa and Argentina.*

*The small but extremely powerful Agritractor of 1928 (above right) and the Sun tractor of 1930 (below) could easily draw a six-furrow plough in heavy clay. They used the same engine as the Speed-Six wagons. The wheelbase measured just over 7 ft.*

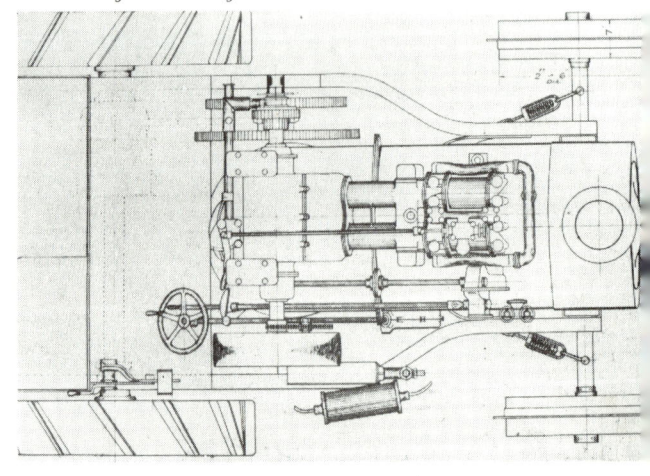

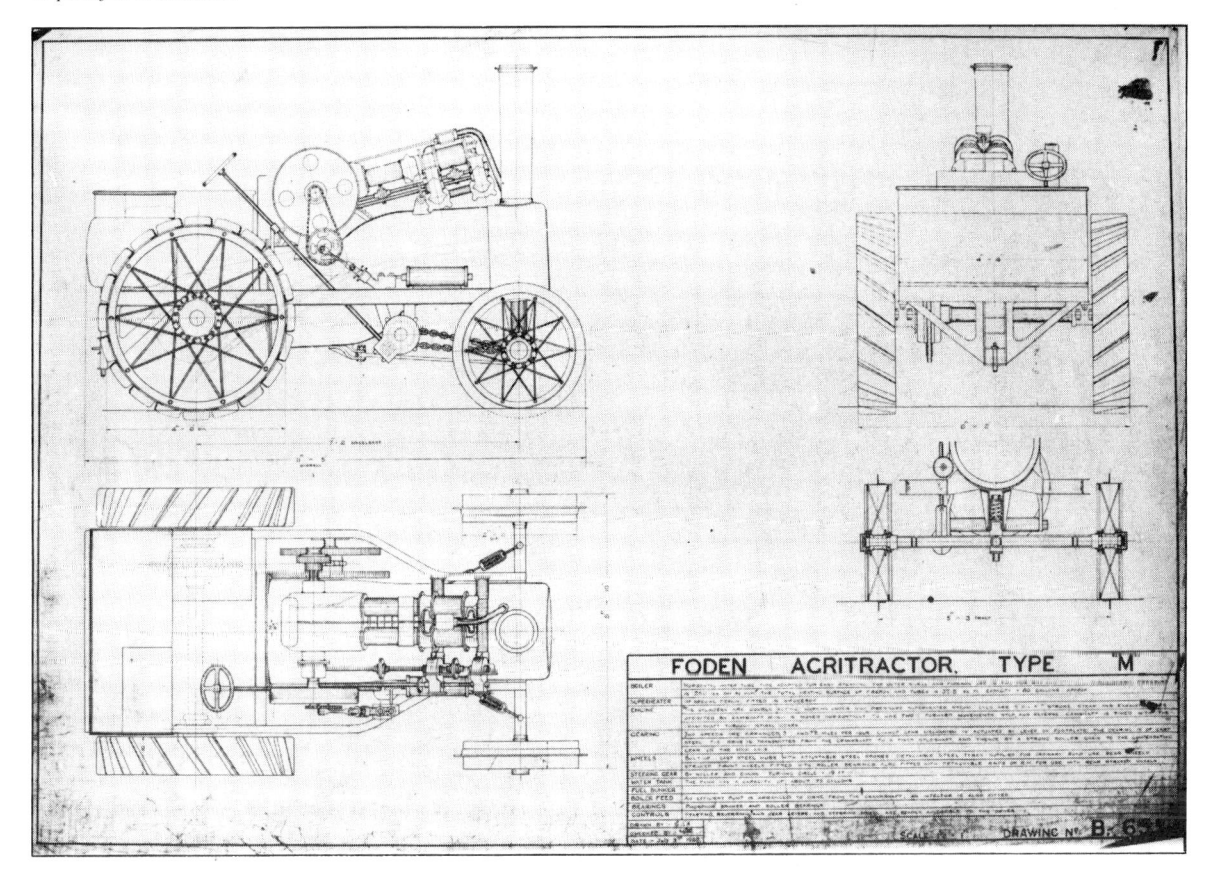

This amounted, in the view of those within the haulage industry, to a licence to pillage and rape, since few if any members of the committee had any road haulage background, whereas many had connections with railways, shipping and land interests.

The Salter Report, as it came to be known, found heavily against existing trends in road haulage. The report claimed that steam wagons were dangerous, presenting a fire risk to property and liable to crush innocent bystanders beneath their wheels. It went on to suggest that no vehicle axle-weight should exceed eight tons loaded, and that any vehicle which weighed over four tons unladen should be heavily taxed. It was widely suspected that the oil companies had done some pretty heavy lobbying in order to achieve these heavily biased answers, but the steam wagon industry could prove nothing.

When the provisions of the Salter Report were incorporated more or less as they stood, and without Parliamentary debate, into the Road and Rail Traffic Act of 1933, the fury of the industry was limitless, but to no avail. The annual tax on a steam wagon was now as much as £235, like it or not. The new legislation had sounded the death knell of the steam wagon, and with it that of those firms exclusively dealing in such products. Among the firms that immediately went out of business were Clayton & Shuttleworth, Mann, Atkinson, Robey, and Allchin, while Fodens staggered under the blow, and Sentinel retired to lick their wounds. Sentinel eventually reappeared with a range of lightweight steam wagons of astonishing refinement, and continued in business for several years more. Atkinson eventually re-formed, but Fodens finally abandoned the manufacture of steamers, delivering the last one overseas in 1934.

Suddenly the need for those diesel vehicles that

E.R. and his old friends in the company had been developing was obvious, even to those who had been blind a year or two earlier. But, as far as E.R. was concerned, their cries for his help had come too late. At the end of 1932, at the age of 62, he had grown tired of all the bitter wranglings, the senseless dismissals of good men, the bad management, and the waste of the few resources the company still possessed, and he retired, to live in relative peace in Blackpool.

With E.R. gone, and many of the men who had worked with him on the diesel trucks dismissed for some reason or other, there was precious little on which to build a new product range. The total number of employees was down to about half of what it had been in the mid-'20s — little more than 200 men. Trading was catastrophically bad in 1933, and for the second year running the balance sheet showed huge losses. The loss on trading was bigger than ever at £53,731, and with various other items like investment losses, the grand total was £72,526, while the bank situation was scarcely less happy with an overdraft of £46,224.

This situation was blamed entirely on outside forces in the annual report, which read: 'We have been, and are going through very trying times. The change over from Steam Wagons to Diesel and Petrol means a good deal of expense. It is true that twelve months ago I mentioned the same thing, but the position has been made very much worse since then by the publication of the Salter Report followed up by a new scale of taxation imposed by the last Budget. The result of this has been, or will be I am afraid, to throw off almost entirely the Steam Wagon. It is only a little over three years since we brought out our latest type of Steamer at considerable expense, and this is practically obsolete owing to the Salter Report and the Budget, and the jigs and effects and materials which we have made and stocked for these models have to be treated as scrap, and our customers who have bought these wagons have got before them a very serious problem, and I am afraid are faced with losses which were unexpected. The Salter Report was a very drastic one and we can only meet it by treating our Steam Wagon Assets in a very drastic way too, and I do ask you as Shareholders to look at this from our point of view as Directors, and not

*The R-type (below right) and S-type chassis (above), built in 1932 and 1933 were in effect development stages between the prototype diesel and the long-running DG models. The R in particular was relatively heavy at 4.9 tons, especially as tax concessions were available for vehicles weighing under 4 tons.*

blame us. Scarcely was our Annual Meeting last year over, but this Report was issued. Up to then we had before us a very hopeful year, but immediately the Report came out the sky was clouded at once. In some cases orders were cancelled, everyone was uncertain as to what the outcome would be and held back until they had really considered their position. This went on until the Budget, and while that struck a severe blow at road traffic, it did at any rate, remove some uncertainty and left the road haulage people with some feeling that they then knew the worst, and trade has, in consequence, gradually been picking up since then. We are not out of our difficulties yet with regard to the Steam Wagon, as right and left we are faced by old customers who wish to exchange them for the new Diesel.

'As to our new types of vehicle, during the year our Drawing Office Staff has been working at high pressure, and our range of vehicles has been extended, so that now we can offer a customer vehicles with carrying capacity of from 30 cwt, to 12 tons, the heavier ones being Diesel and the lighter Petrol driven. A new and patented design of trailer has been made and sales of this have commenced satisfactorily. We are also developing a patent pre-selection gear box, and hope to have this on test in a few weeks. A new Six wheeled Trailing Axle Bogie has been designed

*A painful transition*

and after satisfactory tests some sales have been made. We are further looking in the direction of making arrangements for the maintenance of public service vehicles and are expecting to exhibit this and other designs at the Olympia Show in November, in fact we have already sold the chassis of our first long distance passenger coach of all-metal construction.

'Our Boiler Shop which was kept going by the Steam Wagon trade is occupying our attention and we are getting some trade in various forgings, etc., and work in connection with the Superheater Company. *(This was the Whittle Superheater Co. set up by Wood Whittle in Manchester.)*

'We have had a very anxious year, as I have already said. We have certainly tried to maintain and extend the trade of the Company, and we are all hopeful that the future will in due time bring its reward, although one must admit that competition today is very keen.'

Nevertheless, there were sufficient men remaining who knew enough about the diesel trucks to launch a steady trickle of them on to the market, and fortunately there were sufficient numbers of loyal customers in the harsh world outside to buy most of what could be made, and slowly some semblance of order returned. So much so, in fact, that despite continued disasters with various petrol-engined trucks, the 1934 balance sheet showed a small trading profit of £13,835 out of a total profit figure accepted at the AGM of £17,084.

Wood Whittle was by that time a board member with responsibility for works management, and the annual report painted a modestly encouraging picture. But optimism was not to last long, and the patched-up economy of the ailing company was soon to part at its tired old seams. While other companies sold their diesel lorries in large numbers to a rapidly recovering outside world — not least the new and thriving company called ERF which E.R. and his son Dennis had established — Fodens could manage no more than a trickle, and internal problems at the Elworth works seemed to grow daily more severe. The transition from steam to diesel, which could have saved the company some years earlier, and indeed should have been saving it then, was still not completed.

Had the rot set in too deeply? Was the decay that had eaten to the very roots of the old company too far gone to be eradicated? The answer, it seemed, was yes. It was 1935, the depression was over, and yet Fodens Ltd looked all set to record the greatest ever loss in its existence.

# Chapter 14

# The return of Mr William

We have seen how William Foden's stabilising influence was missed after he emigrated to Australia. It is impossible to say how Fodens would have fared had he stayed with his brother to keep an eye on things at Elworth, but one thing is certain: the results could scarcely have been worse.

William's son Ted was the first to tire of the vast Australian outback, and by 1929 he was back in Britain working as a freelance salesman. He knew how bad things were at the old factory, and occasionally mentioned it in his letters home, though just how interested they were is hard to assess. A handful of loyal employees, many of whom had had their photographs taken and inserted in a commemoration album presented to William on his departure, decided to send a cry for help to 'Mr Willy', as he was affectionately known, when the situation went from bad to worse in 1934.

William Foden had no intention of returning to manufacturing — he was quite happy farming — but he was planning a European holiday when he received the telegram from his old employees, and decided to make a detour through Sandbach and Elworth to see just how things were. What he found so appalled him that he resolved on the spot to give up his dreams of ending his days as a comfortably off farmer — which he was already well on the way to becoming — and set about restoring the company which still bore his and his father's name to something like a going concern, even if it could never be the same as it was in its prime.

Few men can have returned to such a mess as did Willy Foden, and nobody would have blamed him if he had turned right round and headed back to Australia once the extent of the problem was revealed. The trading year was just ending as Willy returned to the seat of his old triumphs, and a trial balance on the accounts revealed huge losses. The accountants went over the books with more than usual thoroughness, and all manner of indiscretions were uncovered. Mr Willy was working as Managing Director Designate, with no proper board mandate, merely a gentlemen's agreement that his appointment should be raised at the next AGM. So glad were the few remaining old employees to see him back, particularly as his brother E.R. Foden was doing well in competition just up the road, that an immediate air of new confidence began to spread around the works. In fact there was precious little reason for confidence of any sort, with or without Willy Foden. The year's trading had resulted in a massive gross loss of £65,476, almost as great as that of 1933. In the end it was not the worst ever, but the point was that it should not have happened at all. At the AGM that August, the chairman, Francis Poole, gave the hushed shareholders his version, and put the blame squarely on the shoulders of the luckless Wood Whittle. After outlining the financial position Mr Poole went on to explain: 'These three items — writing down of stock, loss upon the light type of vehicle, and heavy allowances to customers — account very largely for the loss which is reported. There is, however, another factor of no small importance. That is the large extent to which experimental work has been undertaken, all of which has been written off to revenue.

'At the same time I am sorry to have to tell you that sales during the year fell by an amount of £66,427 18s 2d as compared with the previous year.

'Our Company has passed, in the last few years, through a stage of transition, and every stage of transition is a difficult one. Our Company built up for itself an international reputation in the manufacture of steam vehicles. Time came when this method of transport had to give place to new and improved ideas; oil displaced steam, and this Company perforce had to adjust itself to the changed conditions. Such an adjustment was bound to be a long and difficult task, and the losses on trading which have had to be reported in recent years — culminating and, I hope, ending with the present year — is a reflection of this fact.

'Moreover, everyone is aware of the difficulties with which the road transport industry as a whole has been confronted during this period. These have necessarily created a certain measure of uncertainty, and in these conditions it has not been easy for any firm to work out a satisfactory manufacturing programme.

'These basic facts should, I think, be borne in mind in considering the position which is placed before you today. Whilst, however, these matters may properly be brought to your notice, I do not wish you to think that I am disguising the fact that the loss which we have to report today has its real root in other and deeper causes. Perhaps these can best be brought to your notice if I refer to the fact that last year Mr. R. Wood Whittle, the General Manager, was confirmed in his election to the Board, and that to-day he no longer holds the position of General Manager. The facts which led to the determination of his appointment will go far to explain the Company's present position.

'About that time Sir Edwin Stockton had been co-opted to the Board. It was largely due to him that certain matters were brought to our notice which led us to believe that the position of the Company required investigation. It did not seem that all was well with the organisation, nor was the position satisfactory with customers and agents. It was therefore decided to have a complete and impartial investigation and examination of the Company's position in all its aspects, and for this purpose a Sub-Committee of inquiry was set up. Of this Committee Sir Edwin Stockton was the chairman, the other members being Mr. Stubbs and myself, and we had the assistance of an independent Secretary and investigator.

*Six-wheeler R-types running at 19 tons gross were developed from the four-wheeled versions in 1934-35. They had a 6L2 Gardner, an eight-speed Foden gearbox and a tandem Kirkstall bogie. Weight was a problem and, even with a simple body, ready for the road these R6s scaled over eight tons.*

*The later S-types were popular with operators, weighing about four tons, capable of carrying eight tons and with good performance and economy from the LW-series Gardner. However, they were expensive to produce.*

'This Sub-Committee of investigation was set up on April 16th last. Its unanimous report, extending to some sixty pages, was available exactly one month later, and was duly signed by each member of the Sub-Committee.

'This Report revealed a number of matters demanding serious attention. It revealed, for example, that the stock figure represented no less than 66% of the paid-up capital and also that excess stocks were being carried and that a large proportion of dead stock was present. It showed, too, that the trend of events was towards a serious reduction in the liquidity position of the Company. It revealed serious weaknesses in the costing system and the existence of many transactions upon which considerable loss had been entailed. It showed the existence of many disputed accounts and of many differences with customers and agents. It showed that heavy losses were being made upon second-hand vehicles taken in part exchange, and that in addition to allowances by way of credit note, guarantee repairs represented a very heavy monthly charge. It revealed serious deficiencies in Sales and Service organisation and the necessity for a complete review of production policy, as well as an entire works reorganisation.

'The Directors felt that such a report should be acted upon without delay. They therefore, by a mutual arrangement, determined the position of Mr. Whittle as General Manager, agreeing to pay him reasonable compensation in consideration of his immediate withdrawal. They then decided that for the ensuing six months the management of the Company's affairs should be vested in the Management Executive Council of three members with an independent consulting Secretary, their actions, of course, being subject to the confirmation of the Board.

'It so happens that at the time when all these difficulties pressed upon us, our old friend Mr. William Foden, who had recently returned to this country, was co-opted to the Board. It was felt that no better person could be found to undertake the chairmanship of the Management Executive Council and to enter upon the task of rebuilding the position of the firm and of regaining and extending its reputation and goodwill, both of which had undoubtedly suffered severely by the various matters to which I have referred.'

Few chairmen can have had such an unenviable task as Francis Poole had on that August day, but he had been with the company a great many years, and he knew when it was time to face the music. Even so it is doubtful if the

*The return of Mr William*

*Above and below In the mid-'30s both single- and double-deck bus chassis were built, using Gardner engines and Foden gearbox and axles.*

shareholders would have accepted the situation as calmly as they did, had it not been for the presence of Mr Willy on the board — for his appointment had been gladly confirmed earlier that day.

Nevertheless, the task facing Willy and his board colleagues was daunting, even to men of their calibre, and the image of Fodens in the transport world had suffered badly. At a time when all the major truck builders in Britain were going from strength to strength, and even the newly formed ERF company was making about 350 heavy trucks a year, accompanied by handsome profits, Fodens were still operating at a huge loss, their products were troublesome and unreliable, and little of the old confidence which they enjoyed among the country's hauliers remained.

With the authority of the managing directorship behind him, William Foden set about clearing up the mess. The first task was to replace most of the old internal accounting systems and this at least gave a reasonably clear picture of what was going on from month to month. The next step was to scrap the majority of the designs that had been produced piecemeal over the last few years, and replace them with a rationalised range of truck designs. This would introduce economy both in manufacture and operation, and enable proper service support to be given to the vehicles in use in the field. The third step was to set up a proper sales organisation that could co-ordinate enquiries with production capacity, so making sure the customer got what he wanted at the right time, and at the right price. Having developed into a successful salesman, Willy's son Ted returned to his old company to look after this side of the operation. Before the 1935 Olympia Show, which took place in November just as Willy was in the throes of tackling his problems, there were but three orders on the books, but the rationalisation gospel was preached during the show, and by the close ten days later, those three orders had grown to 33. Willy and Ted Foden were joined on the Olympia stand by Ted Johnson, who had also returned to active participation in company affairs, and the sight of the three old hands must have restored a certain amount of customer confidence, however ill-founded it may have been. But a sufficient glimmer of hope came from that show at Olympia, and there was enough good news about the results of at least one of the truck models in operator service, to confirm the direction the company should take. This was towards a rationalised model range, a much-simplified sales system, and a strict tightening-up of procedures within the plant at all levels from second-tier management to the shop floor. It was a fairly painful process, judging from reports made at the time, and any person found looking the least bit unoccupied stood in danger of losing his job. Any suggestion of waste or duplication was immediately pounced upon. But such was the esteem in which William Foden was held by the employees — and those who were too young to remember him were soon put straight by the older

*In 1936 a small number of very short wheelbase timber tractors were built with a Gardner engine mounted transversely behind the cab.*

**Above and below right** *The DG-series Foden truck of the '30s and '40s went a long way to re-establishing the company after its near-disaster in the early '30s. They were distinguished by the stylish curved radiator and were powered by Gardner LW-type engines of 4-, 5- or 6-cylinder configurations.*

hands — that a remarkable degree of co-operation was forthcoming, and the in-house reshaping process bore fruit very rapidly. Indeed, by the spring of the following year, a mere eight months after the bad news had been read out to the shareholders, William Foden declared that things inside the works were now proceeding on a businesslike footing. Moreover the revised model range was in course of production, while Ted Foden had succeeded in selling off all the used trade-ins, not at very good prices but for precious cash.

Until then there had been a rather ragged and illogical range of trucks, from the little two-tonner with the low-speed Dorman engine, which for some reason was still being offered despite unbelievable problems in service, to a heavier S-type chassis and the top-of-the-range R-type. All these chassis used different axles, different brakes, different engines ... in fact almost everything was different from one model to the next, and with less than a hundred vehicles a year emerging from the factory gates, the unit costs were astronomical. The design team set about producing a new rationalised truck series that would use a high proportion of common parts whatever the model, yet be capable of accommodating a wide variety of loads and body types. This concept is so familiar in the 1970s that one wonders why it was not always the golden rule of truck building. But although a number of manufacturers had begun to work along those

*The return of Mr William*

lines, including neighbouring ERF in Sandbach, it was a major step forward as far as the old-established works at Elworth was concerned.

The model that emerged from this concept was the DG series, possibly the most famous diesel Foden of all, and certainly one of the toughest. This was just as well, because the company's fortunes were at such a low ebb that one more mistake might have sunk them altogether. But there was no mistake. The DG was a remarkably good truck, and what is more it was available in all sizes and power ranges from a light four-tonner with a three-cylinder 53 hp engine, up to a 12-ton heavy-duty chassis with 112 hp, and numerous versions in between. The Gardner LW engine range was a major aid to rationalisation, because all the major parts of these engines, such as pistons, valves, rings, timing gear and chain, bearings and many other parts, could be used in any engine in the range, from two to eight cylinders. This policy of interchangeability is still pursued by Gardner today, with considerable success. Fodens managed to follow Gardner's example in DG chassis parts, crossmembers, brake systems, axles and cabs. This meant an almost overnight reduction in the problems of manufacturing and stocking parts for service purposes, while at the same time the model range was extended and strengthened. The cost of manufacturing was greatly reduced as well, and the list prices of Fodens came tumbling down quite spectacularly. Lower prices in turn attracted many customers, who had been keeping a careful eye on Foden, perhaps hopeful that the great company would be able to sort out its troubled affairs.

Talk to any old-hand Foden man about DGs and you will hear of their qualities. They were not particularly fast, but were immensely strong, and would stand up to overloads quite happily. They were quiet, handled admirably, had a cab that was both stylish and comfortable by mid-'30s standards, and above all were immensely reliable and durable. They were easy to drive, and although the lighter models were proportionately rather heavy — an inevitable result of rationalised design — they were all very easy to maintain and repair. In short, they were just what many truck operators were looking for. Up and down the country, thousands of the old mid-'20s Leylands, Thornycrofts, Dennises, Albions, and the like, which consumed petrol at an alarming rate and were in any case decidedly outdated, were being scrapped and replaced by modern diesel trucks. The tide of this major replacement programme — the greatest switch in basic concept that the UK heavy-vehicle business had ever known — swelled to a peak during 1937-39, and Fodens were ready for it — just.

Mr Willy's Management Executive Council

was undoubtedly effective. Within a year of the gloomy day when shareholders were told of the firm's serious problems, the chairman, Sir Edwin Stockton, was able to tell his shareholders that the recovery was taking place. They were back in the black, but only just. In view of the fact that that trading year was well under way before the extent of the earlier disaster was apparent, and of the time that must have been necessary to take action, it is astonishing that the affairs of the company could have been re-routed so quickly. But for the year ending April 1936 a trading profit of £468 18s 3d was shown. It was not much on which to pay a dividend, but it showed that the tide had turned. In fact, with various investment income, a gross profit of £2,798 8s 1d had been recorded, and the shareholders seemed happier. More to the point the workforce were happier. They had borne the brunt of the stringent efforts to knock the company back into shape: it had been a difficult year, but the indications were that they were now heading for better times, and there was an air of optimism and confidence in the factory that had been absent for a decade or more.

Despite a sharp increase in the excise duty on fuel-oil — the fuel on which diesel trucks ran — the switch to diesel power continued unabated.

The fuel consumption of the Gardner engines, which had in about five years established a reputation for being the most durable and economical diesel engines available for trucks, at least halved that of equivalent-output petrol engines, and even with the new tax fuel oil was still cheaper per gallon than petrol. This meant that diesel fuel costs were about 40 per cent of those for petrol lorries. Against this the higher initial cost of the diesel truck had to be taken into account, but the heavy end of the UK haulage industry preferred diesels to petrol lorries. That winter of 1936-37 saw activity at Elworth the likes of which had not been witnessed since before Mr Willy left for Australia. New tools and plant were installed, new staff and skilled workers were recruited, production figures soared. Fodens Ltd was back in business as one of the country's leading commercial vehicle builders. That year the trading profits rose to £46,383 gross, or £45,191 net. Compared with the golden years a quarter of a century earlier, when close to a quarter of a million was the profit total, it was not spectacular. But it was enough. From that date there was steady growth and finally Fodens was secure. Mr Willy was happy.

*The first experimental dumper was built on an old S-type chassis, with a primitive cab facing the 'wrong' way.*

*The original highly successful DG-series diesel trucks had a Gardner LW-series engine under an angular coach-built cab, with the gearbox mounted midway along the chassis. Detachable-rim 24-in wheels were fitted to most versions.*

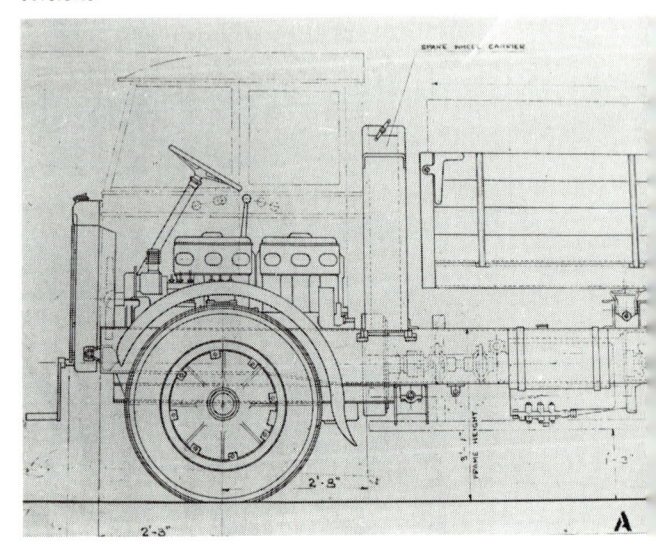

# Chapter 15

# War and peace

As the company emerged from its 1935-37 internal crisis, it became clear that a crisis of a very different nature was looming closer by the week: it was to affect not only Fodens Ltd, but also a large proportion of the civilised world. War was imminent and, though many in Britain hoped it would not materialise, the prospect had to be faced. Even as early as the winter of 1937, it was an odds-on chance that the political manoeuvrings in central Europe would end in a massive confrontation. From an industrial point of view, war was a mixed blessing. There would be the inevitable suffering, privation, even death for many of those involved, but it would also mean an increase in demand for all kinds of engineering products, and that in turn meant secure employment for many. There was bound to be a backlash when hostilities ceased, of course, but good measurement could cope with that as it had done after the 1914-18 war.

In the factory, which was increasingly busy during those years before the '39-45 war, thoughts were generally very far from the looming conflict. The ragged production of the overweight R series with its heavy and troublesome eight-speed gearbox and massive L2 engine had been left behind. Even the more successful S-type with its more manageable Gardner LW type engine and much lighter weight had been discontinued in favour of the rationalised DG series trucks, which were finding good sales among customers throughout the country.

To replace the smallest trucks (the two-tonners which had expired in a plethora of breakdowns, failures and disasters) a new lightweight truck had been designed, and by 1937 it was in full production. This was the OG4-type, powered by the splendid new 4LK Gardner engine, which developed a healthy 60 hp from 3.8 litres, and was remarkably smooth-running for a four-cylinder diesel. So light were the OG4s that, even with a body and hydraulic tipping gear, or a coachbuilt van body, they went on the road weighing under three tons. This meant that they could operate legally at 30 mph, which put Fodens in close competition with machines like the Leyland Lynx and Cub chassis, and the new Seddon, which had been developed not far away in Salford.

Meanwhile the DG chassis went from strength to strength, and the workforce slowly grew until by the beginning of the war it totalled around 1,000 people. It was a remarkable transition from the bad years of the mid-'30s. The company was making good profits, the employees were earning good money. Specialised trucks were beginning to appear, such as a heavy tractor capable of hauling as much as 50 or 60 tons, and (for export use) heavy-duty tippers, which in the course of time became known as dumpers.

The war began quietly enough up there in rural Cheshire. The late-summer weather of 1939 was particularly fine, and folks tended their gardens much as they always did at that time of year. Serious-looking chaps came round and talked about air-raid precautions, and some of the menfolk were pressed into becoming something called an ARP warden. Gas masks were issued, first to the school-children, then to everyone else. The war seemed very far away, and unconnected with the tight little community that surrounded the Foden works. But gradually there was an increasing stream of strangers visiting the factory, with bowler hats and briefcases, not at all the sort of thing that local folks carried with them on their daily business. William Foden had long had associations with the military authorities, beginning with the War Office Trials nearly 40 years previously and continuing with the supply of wagons and munitions during that earlier war. It was almost a foregone conclusion that 'they' would be in touch with Mr Willy yet again, and so they were. Long conferences went on behind locked doors in the Elworth works, and in London too.

It did not take long to work out what Foden's rôle was to be in the conflict; this war would be won with the aid of mechanised transport. Those that were not fully mobile would suffer grave disadvantages and so the rugged and tough DG trucks were to be pressed into service as heavy transports for the Service Corps, the Ordnance Corps, and later for REME.

Armoured strength would also be a decisive factor, it was calculated, and in that sphere of engineering, Fodens could make their contribution. The prewar trucks had started out with a high proportion of proprietary components, but Mr Willy's policy was to switch to in-house manufacture as far as possible, and there was a great deal of new machining capacity for all kinds of components, and considerable scope in specialised metal fabrications too. Could Fodens develop and build tanks? Yes they could, it was decided. During the 1914-18 war, Fodens had made many thousands of heavy shells, but this time it was thought that the emphasis would be on lighter armaments. Fodens were asked to tool up to make 20-mm cannon shells for use in aircraft (such as the later marks of Spitfire and Hurricane) and in innumerable defence installations on buildings, vehicles and above all ships, including the vital merchant ships that were to keep the country alive during the following six years.

The production of those shells raised all sorts of problems. For a start, there had been few female workers at Elworth since the previous war, except in the offices, and when large numbers of local girls began training for their

*The OG4, even with a coachbuilt van body, weighed under three tons and had a gross weight of eight tons. These models competed with vehicles such as the Leyland Lynx and the new Seddon in the '30 mph' light-truck class in the immediate pre-war years.*

new wartime jobs, the provision of toilet facilities produced an immediate domestic crisis. The solution of that problem was relatively easy once someone applied a bit of thought, and bricks and mortar was all that was needed. But shell production running into thousands, and eventually into millions, was a different kettle of fish.

Fodens Ltd, by its very nature, was unaccustomed to the intricacies of mass-production, despite the rationalisation that had taken place to boost the output of the DG trucks. A dozen trucks a week was one thing; umpteen thousand shells a week was something else, and weird and wonderful were some of the production engineering department's creations for boosting that production. A cannon shell is not a complex device. It has a case pressed and rolled from sheet metal, and a hollow heavy 'shot' portion which emerges from the gun barrel to strike the target and explode. The detonator capsules, supplied from outside, were fitted into the base of the casing, and the shells then went to an Ordnance factory for filling with explosives. The work needed a large number of relatively small machines, whereas Fodens possessed a relatively small number of large machines. But multi-spindle jigs and tools were devised to achieve six or more operations at a time instead of one, and slowly they got into the swing of it. Shell production did not really get under way

*War and peace*

until 1941, nearly two years into the war, and even then the July total was an uninspiring 6,000 shells. One raid on one sea convoy could swallow double that quantity in about 20 minutes. However, such were the efforts of those girls, and so ingenious the production men in developing new ideas almost daily to step up the output, that before the end of the year 60,000 shells a month were coming out of Elworth, and this rate continued almost until the end of the war. They worked three eight-hour shifts a day, and the total output from 1941 to early 1944 was 7½ million shells.

If shell production was a bit slow getting off the ground, despite its eventual magnitude, tank production was another story altogether. A tank is essentially a very different machine from a truck, bar the common factors that both have an engine to make them move, and running gear of one kind or another on which to roll along. Construction techniques, pure engineering philosophies on fits, tolerances and target life, operational considerations like costs, economy and weight, are all totally different, the main consideration being that with a tank survival is paramount, and if you cannot achieve that, the cost of repairs and running are of little consequence. It was not until Christmas 1939 that Fodens' engineers first came to grips with the problems of making a tank. The Crusader was to be Fodens' chosen part of the war effort, and the Ministry of War Production teamed them up with part of the Nuffield group and the Ruston diesel and construction machinery company under the code name 'Nuffield Mechanisation Aero Group'. There was precious little to do with aircraft, although a few aircraft parts were made by the Nuffield part of the group. The main job was tanks. The engine of the Crusader was the American-designed Liberty engine, a big petrol unit with lots of low-speed torque. These were built in Britain by Ruston's and were transported to Elworth by road. Some small parts came from the Nuffield wing, but transmission, hull, undercarriage and armament mountings were Foden responsibilities, as was the overall design and development at that stage. A whole range of new skills had to be learned, not least those of machining and welding armour-plate steel. A lot of detail design work was involved, too, because the overall design concept left a lot of 'how-to' questions unanswered.

That year of 1940 brought home the fact that this was to be a tough war. The remaining months of 1939 and the spring of 1940 had been relatively quiet — a period which became known as the 'phoney war' — but then Holland, Luxembourg and Belgium were overrun in May, and the German forces pushed into France. By June Paris had fallen, and the British Expeditionary Force in France was literally toppled into the sea at Dunkirk, where it lost all its equipment, and was rescued by a 'yachtsman's navy' of small boats. August and September saw the Battle of Britain being fought for supremacy of the skies, and further afield the Axis forces spread the war to the Middle East and the Balkans. Despite the immortal Churchillian broadcasts to the nation, nobody could disguise the fact that at that stage the war was not going well. Undoubtedly this knowledge boosted efforts at Elworth, as it did in countless other industrial enterprises, and such was the remarkable progress in the learning of new skills and techniques, that less than nine months after the commencement of tank work, the first Crusader rumbled out of the back entrance to the works and set off down the hitherto quiet Cheshire country lanes on its first test run. To the delight and, it must be admitted, barely disguised surprise of some of those involved, it not only moved under its own power but worked very well indeed, and after a few running adjustments it was delivered to the War Office at Farnborough on October 31 1940.

The following night enemy aircraft all but obliterated the city of Coventry, heart of Britain's motor industry and, of course, the scene of a great deal of military manufacture of various kinds. The war was getting close to home. Somewhat hurriedly a team of men with ladders and paintbrushes turned up at Fodens and proceeded to slosh different colours of paint all over the buildings. It was called camouflage, and although it was a hopeless job to try and conceal a relatively large factory lying right alongside the main London-Scotland railway line from air-reconnaissance cameras or bomb-aimers' sights, there was some chance that the job might make specific areas of the works hard to identify. In any case it was good for the morale of those who worked inside, and so the amazing paint job was completed, with farm buildings painted on big walls, blotches of green and brown on roof spreads, and all manner of abstract greys and browns in between.

In the following months a steady stream of tanks rumbled out of the gates, Crusaders in the first instance, followed by the heavier Centaurs, with thicker armour but similar guns. In all, 770 heavy tanks were produced, no mean contribution towards the struggle that involved the very survival of every family in Britain. By coincidence, tanks were the main war product of the Leyland Motors Ltd plant about 40 miles away. Leyland had been a bitter competitor of Fodens in peacetime, but this was war, and there was considerable co-operation between the two on various problems associated with tank production. Leyland made Comet tanks, faster and more mobile but lighter-armoured than the Centaurs, though of course the engineering problems were very similar.

Fodens were, however, truckmakers at heart, and truckmakers they continued to be during that hectic period of hostilities. Road transport was a vital strategic requirement both in the field of war and at home. Industry was working as a highly integrated team to produce the hardware of warfare, and clearly it was unwise to allow any one factory to be the sole supplier of any item, for the simple reason that a single well-placed bomb could easily dry up the total supply: so production was spread. Once the initial 150 Elworth-made tanks were completed their component manufacture was spread over a wider number of firms throughout the Midlands.

To bring all these bits and pieces together meant a sophisticated transport network, and although the railways did much of the heavy work with bulk supplies of fuel, ore and steels, it was the road transport industry that provided the invisible links between the hundreds — even thousands — of factories working on war contracts. This job was done by fleets of heavy trucks owned by numerous industrial concerns and hauliers, organised by the Ministry of War Transport to fetch and carry whatever was needed, wherever it was needed. This meant many miles through all weathers, air-raids, fuel shortages, and all the difficulties of a nation engaged in the throes of a life-or-death struggle. Among these fleets were no less than 1,018 heavy

*Foden heavy-transport trucks formed the backbone of many RASC companies during World War 2. They were based on the pre-war DG 12-tonner.*

*A convoy of Fodens goes off to war, exactly as they did in the previous conflict. Compare this photograph with the one in Chapter 6.*

Fodens, mainly DG models, but including a sprinkling of S-types and even a few Rs. On short haul jobs, many old steamers were brought out of their showman's yards — which is where most of them had gathered — and were used for heavy haulage of big castings and forgings. For this purpose the restrictions which the 1933 Road Traffic Act had placed on the use of steamers were temporarily suspended, and many an old C-type and Speed Six enjoyed the freedom of the highway once more.

There was equally vital transport work to be done in the actual theatres of war. Transport was supplied to the fighting regiments largely by the Royal Army Service Corps, which operated heavy six-wheelers, mainly of Foden or Leyland manufacture, while the regiments themselves had their own lighter trucks. The job of the RASC was to move heavy stores — anything from cooking stoves to building materials. The Royal Army Ordnance Corps had its own heavy truck fleets for the transport of weapons and munitions, while the Royal Engineers had their own fleet for transporting things like portable bridging gear, earthmoving equipment, demolition gear and cranes for handling stores. Firms like Bedford, Austin and Morris supplied the light trucks for the regiments, and famous vehicles like the QL Bedford and the Austin K2 and K3 toiled in their thousands throughout Europe. The heavy brigade supported them too, bringing up their rations, ammunition and weaponry.

From 1940 to 1944 Fodens supplied 1,750 heavy trucks to the War Office, all DG models and most of them on three axles with a design gross weight of 19 tons, although they were frequently used in excess of that somewhat arbitrary figure. The military DG was almost identical to the civilian version that had been developed in the last few years before the war. The main visible difference was that it ran on large-section track-grip tyres, size 13.50 × 20, instead of the popular 40 × 8s of its civilian counterpart. There were other modifications, such as hefty radiator-guards in front of the cab, heavy towing-hooks front and rear, gas shields below the windscreen, and stowage for the specialist bits and pieces of war like sand-tracking, weapons, and spare fuel cans. Oddly enough none of the DGs was built with a gun mount in the cab roof, which most military trucks had at that time. Apparently this was because they were officially non-combatant vehicles, but events soon changed that, and subsequently many DGs sported roof gun-mounts.

The DG Foden was powered by the Gardner 6LW direct-injection diesel, which developed 112 hp, although some of the four-wheeled versions had the five-cylinder 5LW version of the Gardner. Transmission was a five-speed constant mesh design, of which there were two types. Both had the same basic four-speed main box, but there was one type with a step-up or overdrive fifth gear for use in high speed applications, and

*Above Fodens in an MT park in North Africa in 1942 made an impressive sight. These have been fitted with cab canopies to keep the roof panel cool, and they were also employed to carry items such as camouflage nets.*

*Below Although most of the War Office production comprised heavy six-wheeler transports, about 200 lighter four-wheelers were built, basically DG4/7s in military trim.*

a second type with a direct-drive top gear but a 'super-low' gear, or crawler, for giving high tractive efforts in difficult conditions. Most of the military DGs had this 'super-low' five-speed gearbox, which was wholly designed and manufactured by Fodens, due to Mr Willy's policy of making everything at Elworth in order to control quality. Drive axles by that time were Foden-made as well, with the overhead worm-drive that is still a feature of multi-axle Foden chassis today. The cab was a very strong coachbuilt type using what was called 'composite' construction, which meant that the basic timber structure was reinforced with steel in critical areas, and panelled in thin sheet steel.

One of the first units to be re-equipped with DG Foden six-wheelers was an RASC company based at Baghdad in Iraq, and the trucks were shipped out there via the Cape and the Gulf in mid-1941, to support defences of oil installations which were threatened by both the German forces from south-east Europe and the Italians in Africa. As the war progressed, the trucks were driven across the desert tracks into Jerusalem,

*War and peace*

then down to the Suez Canal, across North Africa and back and eventually up through Sicily and Italy into mainland Europe. Very few of them dropped by the wayside either through enemy action or mechanical failure, although most of them were very battered and war-scarred. Less fortunate was another fleet which worked in Greece supporting that ill-fated campaign. When the Allied forces were finally ejected in April 1941, there was no means of shipping out such heavy items as trucks, and even the men themselves were hard pressed to get away in assorted light ships and destroyers of the Royal Navy, to fight on in Crete. Those grand Fodens, so sorely needed elsewhere, were simply driven over the highest cliffs that could be found, and were smashed to pieces on the rocks below, to prevent them falling into enemy hands. Such are the fortunes — or misfortunes — of war.

In the later stages of the war the Foden six-wheelers and others from Leyland, ERF, and AEC, joined eventually by American trucks, carried the bulk of the millions of tons of stores needed for the campaign across Europe, while the Normandy landings were accomplished with the aid of portable harbour-sections made at Elworth and nick-named 'Whales', although nobody at the factory had the slightest idea what it was they were making. More identifiable was a series of armoured assault vehicles made for the Royal Engineers to support those June 1944 landings. Tanks were stripped of their heavier armour and main guns and were fitted out as armoured bulldozers, bridge-support vehicles, 'flail' mine clearers, and other specialised vehicles vital to progress in an attacking situation. Here again, few at Elworth knew the precise purpose of what they were making; they were simply presented with a specification and they got on with producing the required machine.

As the armies fought on across France, into Belgium, through the Ardennes, and on to the Rhine, the heavy transport went with them. A major feature of the successful land campaigns fought in World War 2 was the excellent logistics support supplied by the service wings of the army, which must have owed a considerable debt to the strength of the British commercial-vehicle industry. In the spring of 1945, those Foden trucks were closely followed by the Foden band,

*A total of 770 Crusader and Centaur tanks was built at Elworth during World War 2. This is a Centaur I; the Crusader was, in effect, a lighter version.*

who were touring the fighting areas giving troop concerts organised by ENSA, and the band ended up with a concert in Brussels on the very day that the cessation of fighting in Europe was officially announced — May 8 1945.

Back at the factory, all was far from quiet. A total of 349 of Fodens' workforce went off to war, and 13 of those were killed in action during more than five years of fighting. In addition to turning out trucks, tanks and shells, the men left

*In addition to fighting tanks, Fodens built almost 100 AVREs (Armoured Vehicles Royal Engineers) used for mine-clearance, recovery, bridge-laying, bulldozing and similar tasks.*

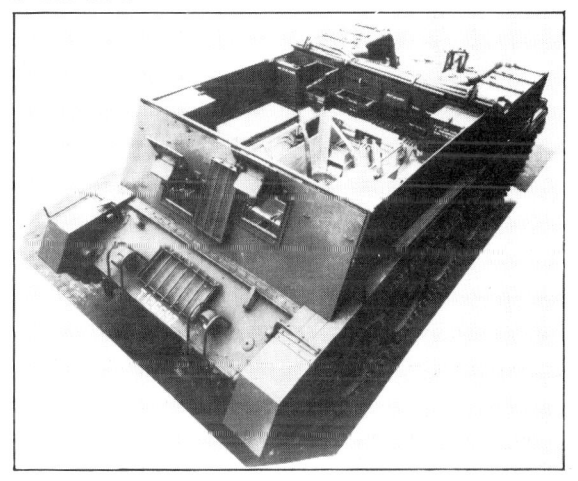

at the factory were involved in other things too. There was a curious 'army' of part-time soldiers in Britain during the war, in addition to the full-time men who fought all over the world, and they were called the 'Home Guard'. The HG was the subject of a certain amount of good-humoured ribbing, but, despite their modest beginnings with armbands as their only uniforms, and staves for weapons at the time of Dunkirk, they grew into a well-disciplined, well-equipped, and useful force. Had an enemy invasion ever taken place, the HG's detailed local knowledge in conjunction with the fighting power of regular troops would have made a very formidable combination. There was a special Foden company of the HG, which rejoiced in the official name of S Company, 11th Battalion Cheshire Home Guard (often called Fodens' Fusiliers), and it numbered more than 80 men at its peak. S Company comprised mainly those who were too young to go to war, those whose jobs made them essential civilians, and those who were too old, many of whom had fought in the previous war.

In addition to the Home Guard, there were several other organisations. There was a works fire-fighting brigade which, although it fortunately never had to work on its own ground, was called to Manchester to help fight fires during the bombing-raids on the city. There was a heavy-rescue squad, which was also called to Manchester to help save people from wrecked buildings, and which also took charge of rescue operations when a power station was bombed just a few miles from Elworth. There was a specialist gas-cleansing team made up of both male and female employees, a general-purpose decontamination squad, several medical and first-aid teams, and last but by no means least a team of stalwart ladies who manned the canteens on both a regular and an emergency basis.

Fodens sponsored a hostel where travelling soldiers and other war-personnel could sleep and eat on their travels, and this became known as the Hotel Foden. Grand though it sounded it was rather basic, but at two shillings a night for a bed and between 7d and 11d for a meal, nobody was complaining. At one period in early 1944 there were large numbers of American servicemen stationed in the area, prior to the Normandy invasion, and Hotel Foden, together with the

*A new cab design appeared in prototype in the winter of 1947-48. It was eventually made in two production versions — the S.18 and a lighter version, the S.19. This was the first new cab design following the long-lived coachbuilt types introduced on the DG models.*

large numbers of pretty girls who worked at Fodens, was a centre of attraction in an area of Britain not really noted for its wealth of entertainment facilities. Quite a few of those Foden girls eventually went to the United States as brides of the American visitors.

The war seemed to finish quite suddenly. As late as March and April 1945, Allied troops were still struggling to force a crossing of the Rhine in sufficient numbers to be of military use once on the other side. Then it all happened very quickly: the breakthrough came both in the west and the east, and by the beginning of May it was 'all over bar the shouting'. Although the aftermath of war brought about as many problems to politicians and statesmen as had the war itself, for those not directly involved, life fairly quickly returned to something like its normal self. It took quite a long time — more than a year in some instances — for the men who had gone away to war to come home again, and of course there were those who did not return at all.

Fodens Ltd was fortunate in that much of its war production was virtually identical to its peacetime products, and so it was a relatively simple matter to switch over to brightly painted trucks for hauliers from the matt-khaki-coloured

*War and peace*

army machines: it took longer to switch back from tank production to making components for trucks. The girls who had made the shells went back to looking after their menfolk when they returned home from the war.

The postwar DG trucks were similar in design to their prewar cousins, but had the benefit of numerous detailed improvements that those tough years had brought. Vacuum servo assistance to the hydraulic brakes was now standard on all DGs, and bigger brake drums were fitted. Improvements to the gearbox had been made, especially the five-speed overdrive version which had been none too reliable before the war, but which by 1946 was a thoroughly good unit. The old composite cab was replaced by a new design called the S.18, which dispensed with the external radiator (the first Foden to do so), and instead had a styled grille on the curved front panel hiding a straight tubed radiator underneath. This cab set the fashion for Foden cab-styling for many years, and the basic grille still remains.

During the frenzied activity of the war years, an exciting new project had begun to take shape at Elworth. It began with full approval of the various authorities who governed such activities in wartime factories, and took the shape of an advanced diesel engine with a mechanical supercharger, working on the two-stroke cycle instead of the more usual four-stroke one. With rationalised component design it would lend itself to manufacture in a variety of sizes and power outputs, with all manner of military applications. It would also, it was considered, make a useful vehicle and marine engine for civilian applications. But before it reached any of those stages, there was a great deal of research to be done. Fodens had never built a diesel of any sort, let alone an advanced design such as this. The main engineering team concerned with the new engine consisted of Eddie Twemlow and Jack Mills, both of whom later became board members, with Albert Pierpoint, Bob Johnstone and Harry Mason (a descendant of Edwin Foden's right-hand man Fred Mason) joining the team later. Ted Gibson, who was later to become Service Manager, was also involved, as was E. W. M. Britain.

The early test engines were single-cylinder units, built to develop the combustion process to

*The FE4 two-stroke engine in its Mk 1 version developed 90 hp at 1,700 rpm. This was the first Foden automotive two-stroke, and was first fitted into a truck in 1947.*

maximum efficiency, and they had an electric blower feeding in through the ports. Early attempts at using air from the factory compressed-air main resulted in halting work in all the other departments due to air shortages! The blower at that stage was off a Delage racing car, and had been borrowed from Mr Raymond Mays, a racing motorist who also operated a Foden truck fleet out of Derby. To experiment with different blower speeds and the corresponding effects on the engine, the Delage unit was powered by an electric motor and, by that rather crude method, a fund of information was built up on rotor speeds, air deliveries, and combustion characteristics. The development of the engine itself went remarkably smoothly, considering that it made more use of light alloys than almost any other engine at that time outside the aircraft industry, and that it was very different from anything made in Britain in the mid-1940s.

The big problem was the blower, or to be more specific the blower rotors. These were mounted on parallel axes, and the lobes of each rotor had to nestle into channels in the mating part as they rotated in unison, maintaining sufficient clearance for lubrication and expansion under heat, but not so much that air could leak past the joint. The path to beating that particular problem was long and arduous, and most of the early rotors were hand-cut on tool-room equipment, and were finished off painstakingly by hand-filing and dressing. Heartbreaking were the times when hundreds of hours' work would go into making prototype rotors, only to have them seize up after a few minutes running. The eventual production rotors consisted of an austenitic steel core with a special alloy rotor profile accurately die-cast into place, finished simply and accurately on a form-milling machine to give a required rotor-to-rotor clearance of a consistent 4 to 5 thousandths of an inch. There were problems with the longer rotors on the larger six-cylinder engines, too, and a further development programme was needed to make them reliable.

At last the engine was complete and the four-cylinder prototype units first ran during the closing weeks of the war years; this did not mean that the engine would not be needed for military applications. Indeed, thousands of Foden two-strokes have been delivered to military authorities over the years, and even in the late 1970s they are still being supplied to the Royal Marines for use in assault craft. Their attraction, apart from light weight and high power, is that they are still the only engines which will withstand the tremendous stresses set up when thrown to the 'full astern' position from 'full ahead' in the event of an assault landing having to be abandoned at the last moment.

As the war years suddenly gave way to peace — of a sort — the thoughts of the engineering team turned to applying their new power unit to a road-going truck. Even here there were problems. Cooling was one, and gearing was another, because the speed range of the two-stroke engine,

*Extremely stylish cabs characterised the two-stroke trucks like this FE4/8. This is an eight-tonner with the four-cylinder supercharged two-stroke Foden power unit. This type of cab was built for Foden by O. G. Bowyer.*

or the FE-types as they became known, was quite different to that of traditional engines such as the Gardner. But the power output was very satisfactory, with something like 90 hp from the small FE4 engine. This engine was first fitted to a truck barely two years after the official cessation of hostilities, and within a further six months it was a production option. By 1950 the six-cylinder unit with about 130 hp was available, and that went into more powerful trucks, mostly eight-wheelers but some six-wheelers, too. It was an expensive engine, and required different skills in handling and maintenance compared with conventional engines. As a result, the FE4/8 truck model, an eight tonner powered by the four-cylinder Foden engine, was withdrawn once the heavier types were available, mainly because it was too costly for its market, and also because many users in the lower weight ranges tend to look on trucks as elaborate wheelbarrows that need nothing so complex as maintenance. But in the heavier ranges the FE6 was a howling success — an apt adjective, since the high-pitched engine note of those machines quickly became a familiar and exciting new form of music on the highways of Britain. The output characteristic of the two-stroke was such that it required a close-ratio gearbox to do it justice, and Foden's existing five-speed box was not in that category. A number of alternative transmission arrangements were tried, including two-speed 'splitter' axles, and an automatic two-speed overdrive auxiliary transmission mounted half-way along the transmission line, as well as a preselector box known as the SSS which only reached the experimental prototype stage. There were six- and seven-speed transmissions too which went into production mainly for the two-stroke engines, but finally it

*Front-engined FP type double-deckers formed the mainstay of a number of municipal fleets, notably in Chester, Warrington and Derby.*

was Fodens' unique twelve-speed gearbox which provided the solution to the two-stroke engine application among many others, and this design has been retained in its basic form right through to the present-day Fodens.

In those immediate postwar years a whole programme of engineering development was initiated, prompted to some extent by the new skills that had been learned while building military hardware. The transmissions were just one part of the programme; another was brakes. Hydraulic brakes had been standard on the older DG trucks, later boosted by vacuum servos, but on the two-strokes there was no easy source of vacuum from the manifold of the engine, so a new energy system had to be devised. This took the form of a hydraulic booster arrangement to the braking system, in which an engine-driven pump circulated oil at a pressure of 1,000-1,200 psi and, as the brake pedal was depressed, so proportionate amounts of that pressure were bled off to actuate the normal hydraulic master-cylinder with considerably more force than a man could generate on his own. The system first appeared on the FE4 trucks, and was later applied to the bus chassis that also appeared after the war. The system worked well, although it needed very careful adjustment when setting up after installation or overhaul, but so effective were these brakes that they continued in use on most Foden chassis until the adoption of air-pressure brake systems in 1956.

Very few buses had been built by anyone during the war years, and many had been destroyed or run to the point of collapse; consequently there was a boom in bus-building from 1946 onwards, which eventually settled down to a steady level only in the mid-'50s. Fodens had built a limited number of buses and coaches in the early diesel days, but basically they were truckmakers. However, this did not stop them from entering the bus field with enthusiasm in 1946, with a range of very advanced designs. The basic design was known as the FP chassis, also referred to as the series 4, which was distinguished by a cruciform-braced main frame unique to Foden buses. There were single- and double-deck versions with both Gardner and Foden engines, and a great many touring coaches were built on the single-deck versions. Double-decker Fodens were sold in quantity to several municipal bus fleets, and the City of Chester and Borough of Warrington fleets were standardised on Foden double-deckers for years, while the City of Derby ran a large number alongside other makes.

The next stage was the development of rear-engined chassis. A rear-engined double-deck model was developed in 1949 and put into limited production by 1950, pre-dating Leyland's Atlantean rear-engined bus by seven years. But it was the rear-engined two-stroke powered coach chassis which really captured the imagination of the passenger-transport industry. This was known as the series 5 chassis and all the famous names in the coachbuilding business, such as Bellhouse Hartwell, Beccles, James Whitson, Yeates, Windover, Harrington, Burlingham, Plaxtons, Santus and others — most of them alas, long since vanished from the scene — queued up for Foden rear-engined coach chassis on which to base their flamboyant operations. The main attraction was that the absence of a front radiator lent greater freedom of styling and in addition the buses were fast, which attracted the tour operators. In all, more than 650 buses and coaches were built during that postwar period.

In service the Foden-engined machines suffered somewhat from cooling problems, the

**Left** *A rear-engined double-deck bus chassis was produced in 1949-50, with either the Foden FD6 engine or the Gardner 5LW. This cruciform braced chassis pre-dated the Leyland Atlantean by at least seven years.*

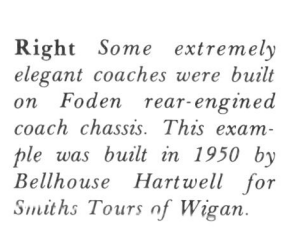

**Right** *Some extremely elegant coaches were built on Foden rear-engined coach chassis. This example was built in 1950 by Bellhouse Hartwell for Smiths Tours of Wigan.*

technique of controlling airflow round the back of the essentially cuboid shape being in its infancy in the late 1940s, and the Gardner-powered version had some favour despite being much slower. Many of those early Foden-powered coaches still exist, however, including the Foden band coach which is eligible for inclusion in historic vehicle events, but still earns its keep.

The difficulty which surpassed all others during those exciting postwar years of new developments and hope, not only at Fodens but elsewhere too, was the acute shortage of materials. Even ordinary steels were rationed until after 1950, while more exotic materials like special alloys were like moondust, and out of the reach of all but a privileged few. These restrictions kept production to disappointingly low levels, and indeed throughout that period Fodens' profit figure on a pound for pound basis, ignoring the changed value, never climbed as high as it was just after World War 1. Nevertheless, there are many who insist that the first years after the '39-45 war were the most exciting and satisfying in the whole recent history of Fodens Ltd. Certainly from an engineering and development point of view this could be said to be true, and there were many envious eyes cast in the direction of Elworth, among them the author's. I was a young apprentice 40 miles away at Leyland Motors at the time, and the sight and sound of Foden coaches, bearing names of what to me were far-away places — Plymouth Co-operative, Salopia, Pride of Bridgewater, and others — and all speeding their spectacular way to Blackpool and other Lancashire resorts, were exciting indeed; they also brought with them the implication that there was a healthy growth in British commercial vehicles — and I was part of that industry.

# Chapter 16

# The postwar expansion

As the first flush of enthusiasm in a postwar world gave way to more profound thoughts of building a long-term future, it became clear that a great deal of new design and engineering work would be necessary at Elworth. The DG range was a good, sound proposition for most normal purposes, but by 1948 it was already 12 years old; and even with the numerous improvements that had been incorporated into the brakes, transmissions and cabs, it was still a truck of the 1930s competing in a 1950s world. The bus chassis at least were modern designs, needing little further development, and they enjoyed reasonably good sales. Obviously a new truck range was needed.

The two-stroke FE series chassis were the first to appear, and to some extent were a disappointment to William Foden, his sons Ted and Reg (who were then on the board), and Eddie Twemlow. It was certainly a handsome truck and worked well, but the characteristics of the new engine meant that, instead of pursuing the policy of 'we make it all', a new phase of using proprietary parts had to be embarked upon. Axles came from either Rubery Owen or Kirkstall, depending on the model, and the standard S.18 cab did not fit too well over the compact Foden engine, so many chassis went out to have cabs built by Bowyers or Duramin, although later a version of the Foden cab called the S.19 was made for those chassis. However, a Foden transmission was fitted; indeed the FE models launched the company on a long and adventurous phase of development in transmissions which was to continue for more than ten years. The two-stroke programme continued into the 1960s as far as trucks were concerned, and still continues in the industrial and marine field, but gradually the demands of legislation and the prohibitive cost of producing a specialised engine in relatively low volume made the two-stroke series too costly for use in normal trucks. However, this did not happen until a long series of improved versions had appeared, including so-called 'dynamic' models, where tuned-length ram pipes in the inlet and exhaust tracts were used to boost the normal input and output of air and gas, increasing the power considerably. Even the little four-cylinder unit eventually developed 120 hp, and larger types up to 225 hp; but that was not until the late 1960s.

Parallel to the development of the FE types with two-stroke engines, more conventional designs were drawn up for those users who avoided advanced engineering like the plague. These designs grew into the FG series trucks and, although they were basically conventional — with Gardner engines and Foden transmission and axles — they were considerably lighter, type for type, than the DGs, had a much better cab and carried more payload. The first FGs came off the production line in 1950 and immediately attracted the attention of the operator customers.

The model range was made up largely as had been the DG, with Gardner LW series engines in four-, five- and six-cylinder versions, and even eight cylinders by 1955, driving to a unit-mounted gearbox and Foden overhead worm axles. Type numbers were built up from the FG prefix, with a number representing the number of cylinders, then a stroke and a second number indicating the

*The postwar expansion*

designed payload. So an FG4/7½ was a 4LW-powered truck with 7.5-ton payload, and an FG6/15 had a 6LW engine and carried 15 tons, which meant it was an eight-wheeler. A variety of gearboxes went into the FG, initially the old five-speed unit in one of its two forms, overdrive or super-low was used, but this was replaced first by a six-speed and subsequently by a seven-speed transmission, both designed and developed at Elworth.

A wide variety of special versions was developed from the FGs: the FGD dumpers for example, and the FGT tractive units and drawbar tractors. So sound was the original design that it stood Foden in good stead for more than a decade, and many of the components and sub-assemblies, in particular the rear bogie of the multi-wheeler chassis, remain virtually unchanged to the present day. They were popular machines with drivers and owners alike, because they were safe, solid machines to drive, with good overall gearing which gave a more than adequate range of performance for that time — remember that the 20-mph speed limit was still in force for trucks over three tons when the FG was introduced — and, as far as the owners were concerned, the economy of the Gardner, and the

*The S.20 cab introduced in 1954 showed the styling trend to come at Fodens, and was fitted to a wide range of chassis. This is an FG5/9.*

durability of the whole vehicle kept running costs low, and vehicle utilisation high.

There was intense competition during those first years of expansion, from ERF up the road in Sandbach, AEC with their excellent Mammoth series, Leyland's Beaver, Hippo and Octopus

*Still retaining the chassis specification of the Gardner 6LW engine, five-speed gearbox, and 40 × 8 wheels and tyres, but showing progress with the new S.18 cab: the FG6/15 eight-wheeler was a real workhorse in many fleets by 1950, including that of the millers, Joseph Rank.*

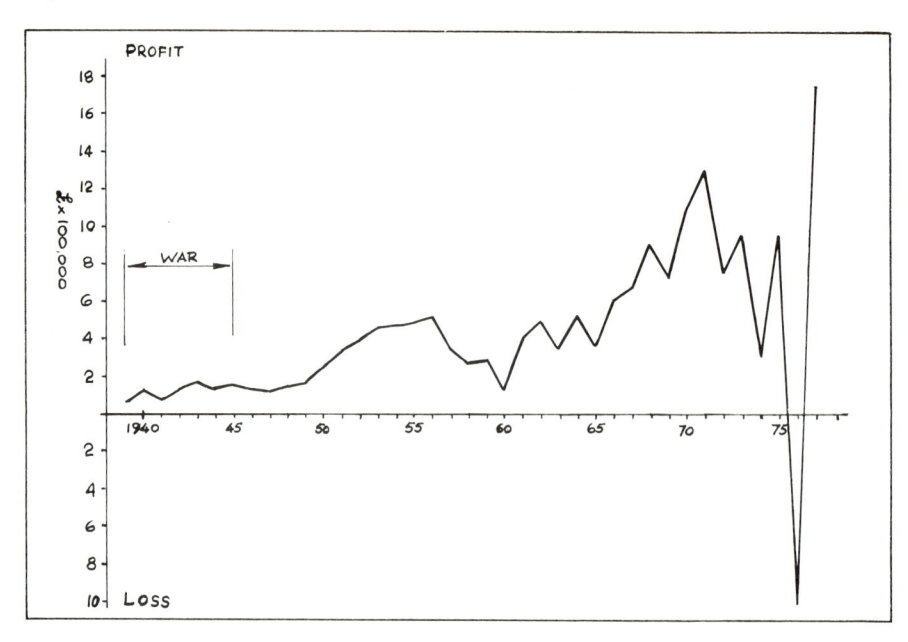

**Above left** *Pre-tax profit figures showed a steady growth of the company after 1940 marred only by the £1.01 million loss in 1976 following the massive plant investment. The stunted growth in the postwar period of material shortages is evident.*

**Above right** *In the Summer of 1954 the motor vessel* Aries *completed a double Atlantic crossing powered by two Foden two-stroke marine engines. David Foden, later to become joint managing director, was one of the crew of four.* Aries *was originally a lifeboat, built in 1929 and later converted to diesel power.*

**Below** *Fodens' seven-speed gearbox was developed in the immediate postwar period and served in most truck types until the advent of the 12-speed unit in the early '60s.*

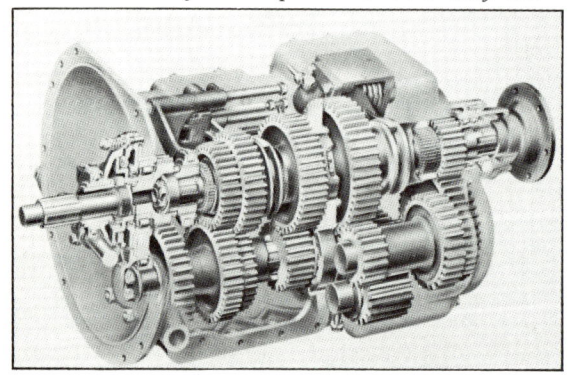

ranges built round the legendary 0.600 engine, Maudslay's Mogul, Atkinson's Gardner-powered range built at Preston, and others including Sentinel which had for so long been an adversary in steam wagons. Even light-truck builders like Guy and Seddon were moving up into the heavier field, thereby providing further competition. But the major threat to expansion was not competition from other makes — that could always be handled, and was basically healthy anyway — but legislation which once again reared its ugly head and dealt another body-blow to road transport.

It all began with highly restrictive legislation in the 1947 Transport Act which required special licences for vehicles travelling more than 25 miles from base. Licences were strictly controlled by the Traffic Commissioners, and the Railways could object to them — which they did at every available opportunity. This was in effect a development of the prewar system but with much more sinister undertones now the state-owned railways were involved. But worse was to come. In 1950 an extra 9d a gallon was added to fuel tax, and purchase tax of 33⅓ per cent became payable on commercial vehicles as well as on cars. The long-haul sector of road transport was nationalised, only to be de-nationalised when the Conservative government of Winston Churchill replaced Clement Attlee's postwar Labour

*The postwar expansion*

administration. However, British Road Services used the name under which the nationalised industry traded, as a state-owned company competing against the private sector. The new BRS spent money freely on large fleets of heavy trucks, notably Leyland Octopuses, and later on built its own trucks to its own designs, using the Bristol Company's facilities and a Leyland engine. One effect of the BRS activity was to provide over-capacity in haulage, and this in turn led to a spate of rate-cutting. The end result of all the political manoeuvring and counter-manoeuvring was that the free-enterprise sector of truck operating and manufacturing had a hard struggle to keep going, and indeed a number of companies went to the wall in the early '50s, including Sentinel, Thornycroft and Vulcan. The depressed level of trading during those years is clearly visible in the graph of Fodens' financial performance, but finally, in 1951, Fodens' profits cracked the £250,000-level, which had been the previous highest in the company history, way back in 1920, and steady improvement was achieved from then on.

The reason for this steady improvement lay not in spectacular successes on the UK truck market, for that was still almost static. Neither was the bus business showing the promise that it had a few years previously. Wisely, as it turned out, the company had decided to diversify its activities. Their new two-stroke diesel engine made an ideal industrial and marine unit, and a promising new field of business emerged from 1951 onwards in generator sets, drill rigs, pumping units, and a hundred and one other applications, while many diverse military and naval uses for the engines were found. Even the royal yacht *Britannia* used Foden two-strokes as auxiliary power units. The heavy-dumper market in emerging African and Asian nations provided a further opportunity for new business, and in the early 1950s the dumper development programme made spectacular strikes.

The first dumper had been made way back in 1937 when, at the request of Hughes Limestone Ltd of Buxton, a second hand S-type chassis was fitted out experimentally as a special site tipper, after a number of accidents when normal road tippers reversed to the edge of high precipices in the quarries. Indeed some trucks had even fallen over the edge, and so the 'new' machine was arranged with the cab in the usual position but with the driver facing the load-carrying part of the vehicle, instead of the other way round. It was a strange-looking affair, but it did the job adequately, and that set Fodens' engineers thinking about heavy-duty tippers, or dumpers as they were subsequently known. Later thoughts abandoned that 'face-the-rear' principle for a variety of reasons, and instead concentrated on providing a really stable, tough platform on which to mount the body. The first real dumper chassis were in effect very short wheelbase

**OVERLEAF**

**Background photograph and left inset** *Early heavy-duty dumpers were beefed-up versions of the FG six-wheeler chassis, with big single tyres on the bogie and cab canopies on the bodies. This type was built from 1951 until the introduction of the half-cab type on a similar chassis in 1954.*

**Centre inset** *A limited number of bonneted dumpers were built for African markets in 1958-60. They were designed to carry 28 tons payload.*

**Right insets** *Once the formula for the heavy-duty dumper had been settled, the overall concept did not change much. These two pictures, taken respectively in 1956 and 1971, show the detail development over 15 years.*

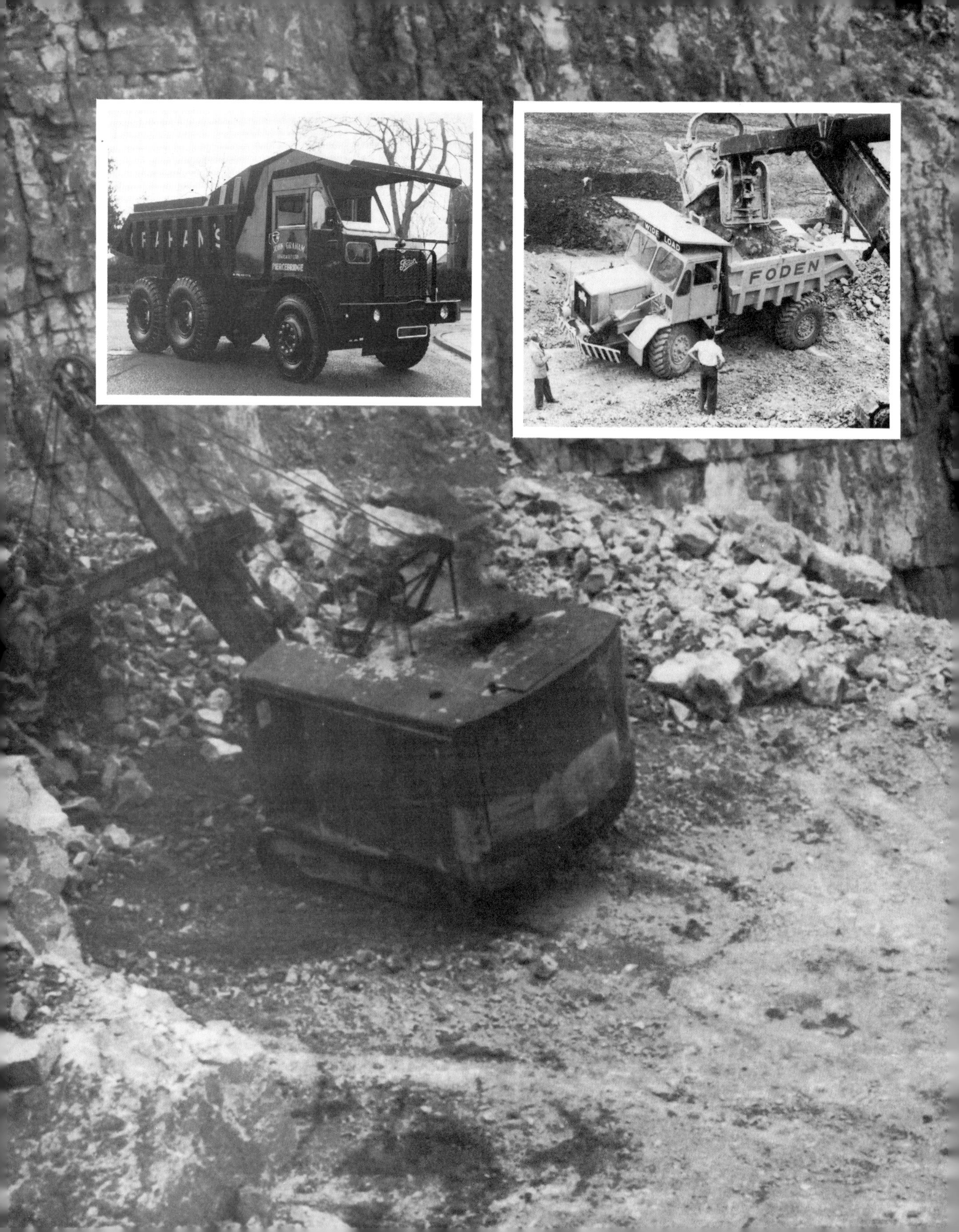
JOHN GRAHAM
PIERCEBRIDGE
WIDE LOAD
FODEN

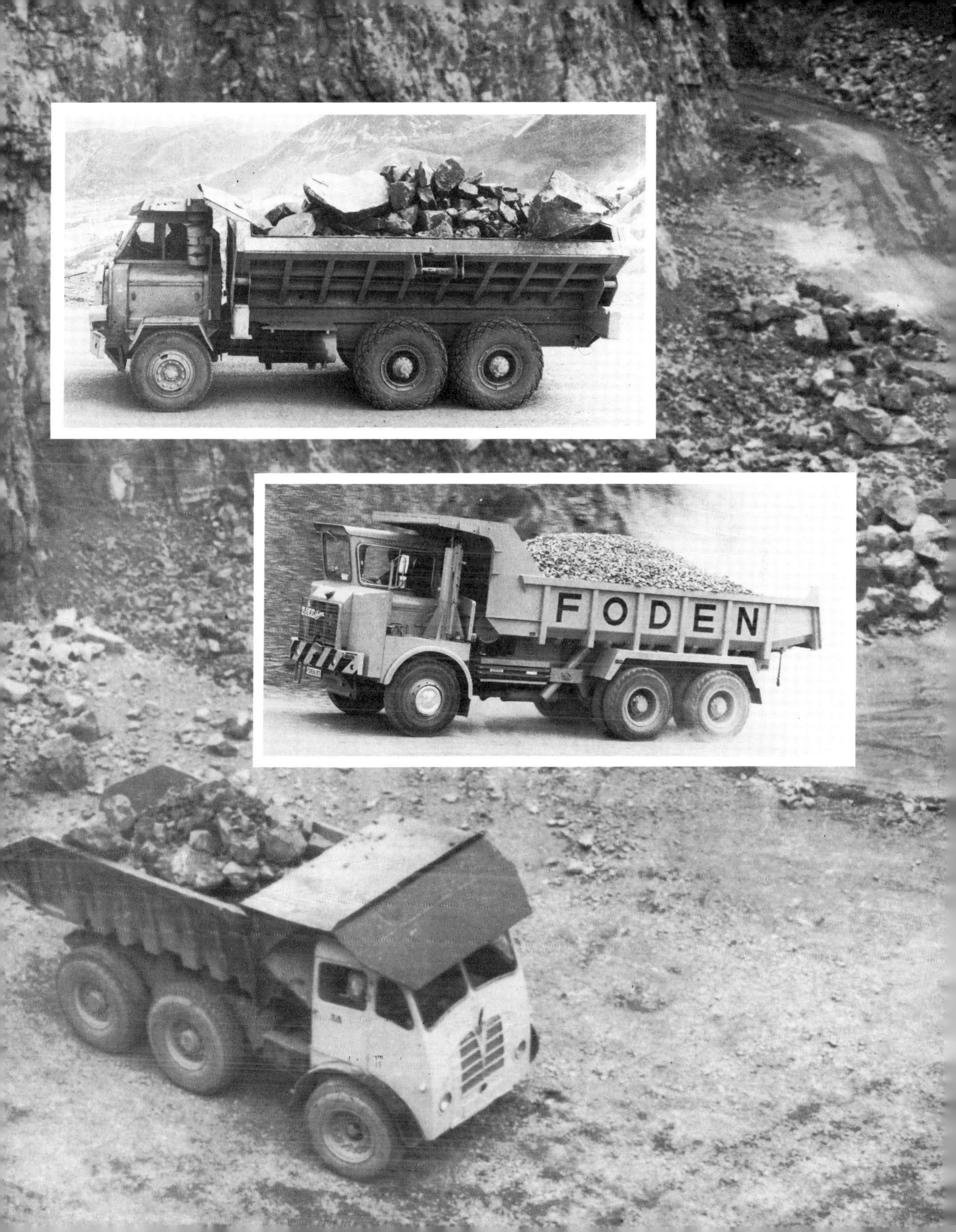
FODEN

six-wheeler DG chassis, stiffened and strengthened, and with a very stiff suspension. The result was a success, and few accidents occurred while they were in use.

The cab was a problem: it got knocked about in the hurly-burly of site work, and in due course a big steel cab shield was added to the front of the body. By 1953 a special steel half-cab had been developed, on the basis that such a dumper never carried a crew of two, so why have a two-seat cab? In any case the half-cab, which was made in heavy-gauge steel, was much tougher than the composite truck cab, and gave the driver better visibility too, an important factor when working on sites and tips where the terrain might be dangerous.

The next step was a two-axle 24-tonner, which was very popular in countless pits and quarries in Britain, although the heavier types based on the same layout, carrying up to 40 tons, were the spearhead of Fodens' export drive for years. In fact the current descendants of those mid-'50s models, the two-axle FC17 and FC20, and the FC27 and FC35 six-wheeler types, represent a steady evolution of the basic designs, resulting in remarkably durable vehicles. (The numbers in the designations represent the carrying capacity.) In between there were some heavier types, built for special purposes but since discontinued because of the rationalisation policy at Elworth, which we shall examine a little later. There was a 28-ton payload bonneted dumper on two axles, introduced in 1958, and most of this model was sold in African markets. Ten years later huge 40-ton and 60-ton dumpers were designed, and built in relatively small quantities, but the manufacture of components of suitable size and strength was uneconomical with soaring wages and material costs, and the super-heavy market was left to Fodens' new trading partners in Europe, Faun Werk, who had a large market in such equipment.

The development of the family of dumpers stood the company in good stead over the years for, as the on-highway truck market went through its inevitable cycles, the construction industry in various parts of the world ordered dumpers at a steady rate. Throughout the 1950s Foden seven-speed gearboxes were fitted, with either Foden diesel or Cummins engines. To begin with the immensely tough Foden double-reduction heavy axle was used, and in later years the equally tough but rather lighter hub-reduction axles were fitted.

Once the supply problems of the late '40s and early '50s had disappeared, and a short period of political and legislative stability had allowed the market to settle down, there was a steady growth

**Left** *Extremely sturdy and stylish Low-line crane chassis were developed in the early '70s using the same basic principles as the 1965 design. This type appeared in 1971 and caused much interest at the Public Works Exhibition at Olympia.*

**Right** *When the first 'spaceship' cab — the S.21 in official terms — appeared in 1957, its flamboyant styling and shape took the truck world by storm, aided and abetted by the impressive-sounding two-stroke engines.*

*The postwar expansion*

*By 1965 a special crane chassis had been developed with a low cab over which the boom was stowed. The engine — in this case a Rolls Royce Eagle — was mounted over the front bogie.*

in both turnover and profits which continued until 1957. Then came the Suez crisis, fuel rationing and more political unrest, and the nation's truck operators decided to 'keep their hands on their ha'penny' for the time being, with the result that the UK truck market slumped, and with it Fodens' turnover and profits. The aftermath of that crisis kept trading at a low ebb for nearly three years, but in the meantime it seemed as good a time as any to update the model range in order to win as much of the market share as possible. The first new cab for nearly eight years appeared in the form of the S.21, more popularly known as the 'spaceship' or 'Sputnik' cab, so called because of its exaggerated curved shape. With the new cab a new uprated version of the Foden engine was introduced, the FD6 MkIII, which developed a healthy 150 hp — more than any of the opposition could muster at the time.

The effect that these new Fodens had on the market was electric. The spectacular cab combined with even more spectacular performance won Fodens back a sizeable share of the market, and the ambition of every truck driver on the road was to have a new Foden S.21. Part of that ambition resulted from the noise made by the MkIII which was even more strident than that of its predecessors.

Despite the undoubted success of the new model range, trading remained poor. Having just recovered from the Suez crisis, the industry found itself facing the threat of re-nationalisation under a new Labour Government; there had been a flurry of wage increases following Suez and the oil-based economic problems that resulted; and there had been a widespread strike of engineering workers which reached almost national propor-

**Left** *Willy Foden breaks the ground for the new office building in 1956, the official centenary year. Among the onlookers are Eddie Twemlow, Ted Foden, Ted Johnson and Bill Foden. Mr Willy was 88 years old at the time.*

**Below right** *In any street full of trucks, the S.39 Foden styling stood out from the crowd. This basic cab shell was used for more than 15 years.*

tions. However, in 1959, the first signs of a real recovery appeared with the withdrawal of purchase tax on commercial vehicles, and the lifting of restrictions on hire-purchase trading. So when, early in 1960, the threatened nationalisation of road transport was scrapped once again, the truck market immediately recovered, and by the end of that year turnover and profits were back at the mid-'50s level. As well as the home market upturn, new export markets were found in such countries as Borneo, Sierra Leone, Kuwait and Ghana, while the long-established South African subsidiary company began showing its first profits for a long time.

From a financial point of view the decade until 1971 showed a steady and strong upward trend both in turnover and profit. Despite the inevitable reactions to periodic credit squeezes and international monetary crises, this was one of the best growth periods in the whole of Fodens' history. The reasons were not hard to find. After the introduction of the FG range of trucks, which were built to a well-rationalised design profile, so many variations and special models were introduced, not least in the big dumper field, that the potential savings in production rationalisation had been largely lost. Steady expansion in various departments within the factory had led to a rather ragged production flow, and a certain imbalance in component supply and stocking was inevitable. Consequently, in 1961, a reorganisation was proposed, both in the production facility at the Elworth works, and in the model line-up. A new chassis-assembly hall was built with two parallel tracks for the volume-built on-highway trucks and tippers, with a third 'specials' line for crane carriers and any non-standard machines that were built, for example military versions and special export machines. At the same time both the design and production of components was reshaped. The new assembly shop had left more space for component production, and new machine tools were installed with the emphasis on proper production flow.

First fruits of this reorganisation were new axle designs, with higher weight capacities to suit expected increases in operating weights and speeds, now that motorways were becoming a reality in Britain. The overhead worm principle was retained, but new ratios were introduced, and the whole drive-axle range was built around two designs, a single-reduction type with weight capacity up to 12 tons, and a hub-reduction type with capacity up to 20 tons. Dimensionally there were many similarities between these axles, so a high degree of interchangeability was possible, which in turn made production and servicing costs much more acceptable. A similar arrangement applied to the front axles, with only two types being shared between all models: tractive units, six- and eight-wheelers, and dumpers. Even this initial stage resulted in 1963 in significant

*The postwar expansion*

cuts in production costs and increases in output. The next stage was even more effective, but before that took place the whole Foden community was to receive a shock of grave proportions.

On June 2 1964 William Foden died. Looking back at that sad event, it cannot have come as a surprise for Mr Willy was 96 years old, and even he was not immortal. But he had been a significant force in the company ever since the days of the great industrial engines and, with the exception of the few years when he became an Australian farmer, there had never been a Fodens Ltd without Mr Willy. It was not only within Fodens that his loss was felt, for William was widely respected throughout the truck world, and on frequent occasions was referred to as 'the father of the British commercial vehicle industry.' Like his father before him he had been a firm, insistent but sympathetic boss to many of the men at Elworth for as long as they could remember, and in many cases for as long as their fathers could remember, too. Few other men can have been actively associated with a company for a span of 81 years, and he had taken that active part right to the last. For the later years of his life he had relinquished the energy-demanding job of managing director jointly to his sons, Ted and Reg Foden, with Eddie Twemlow as engineering director, while he remained as governing director, keeping an eye on the way things were run. It was he who cut the first turf when the new office block and attached workshops were built in 1956 at the time of the company centenary, and it was also he who nodded approval when prototypes of the spectacular new cabs appeared a year or two later. The whole industry — colleagues, rivals, truck operators, drivers, families, employees, in fact everyone connected with trucks in Britain — mourned William Foden. On his last journey his body was carried slowly and gently by the beautiful little three-tonner *Pride of Edwin* which is preserved to this day at Elworth.

But if 1964 was a year of sadness, it was also a year of challenge, for a new set of 'Construction and Use' regulations for trucks operating in Britain was issued by the Ministry of Transport. The main feature of these regulations was that articulated trucks up to 32 tons gross were allowed, provided certain axle-weight and dimension limits were observed, but the rigid eight-wheeler weight limit remained at 24 tons. The result was a big swing in the market towards the articulated tractive unit, instead of the maximum-capacity eight-wheeled rigid truck which had been so popular. A 32-tonner could carry up to 22 tons of payload, while a rigid eight could manage only about 15-16 tons, and this extra margin represented a major revenue boost for operators.

Fodens had their ammunition ready, and a major attraction at the 1964 Commercial Vehicle Show at London's Earls Court was a new Foden tractive unit designed to meet the 32-ton limit. It had a very stylish new cab, the S.34, more spacious than the S.21, with much-improved vision through a large, curved, single-pane windscreen, and with better sound insulation. The cab tilted to give access to the engine, which could be either a two-stroke Foden or the time-honoured Gardner. (Gardners were by that time offering the 180 hp 6LXB in addition to the 150 hp 6LX and the old 120 hp 6LW.) There was a new gearbox, designed and developed by Foden, with 12 close-ratio gears provided by

*1965 was an innovative year. It saw the unique 'Twin-Load' concept, aimed at running up to 32 tons gross using an eight-wheeler and semi-trailer. It also saw the S.34 tilt cab, introduced at the 1964 Commercial Show, in production, making Foden one of the first UK manufacturers of these cabs.*

applying a three-ratio epicyclic splitter box to a conventional four-speed constant-mesh main box. With it went a clutch with sintered iron linings of unprecedented durability, and these two units were eventually standardised throughout the Foden range.

There was also a rather curious device on show at Earls Court called the Twin-Load. This truck was unique; nobody ever tried to build anything like it, although in many ways it was an excellent design. It consisted of a normal eight-wheeler chassis with a coupling perched right at the rear to which was attached a single-axle semi-trailer. The idea was to offer an alternative method of running at 32 tons gross if you did not want to use normal articulated trucks. It also gave the alternative of running as a rigid eight-wheeler when roads were icy or loads were smaller. A number of fleets took up the Twin-Load idea, among them the Harvey's sherry people of Bristol and Colmore Refractories, and in certain applications the combination of a very long loadspace and great stability proved to be good operational features. However, it did not have the interchangeability advantages of the artic concept, and was made only in limited quantities.

The demand for artic tractive units took most manufacturers by surprise; it is estimated that more than 75 per cent of the heavy chassis sold in Britain in the year following the new regulations were 32-ton tractive units. Very few makers had calculated that the swing would be so violent; certainly Fodens did not, and the result was that at Elworth, and at various other factories up and down the country, there was a severe shortage of tractive units, and a corresponding surplus of multi-wheeler chassis. The new rationalisation of engineering product had come just in time, and now was the critical moment to discover whether it really worked, or whether it was a 'paper system' only.

The last stage of the production re-structuring included the installation of a computer to control stock, materials, parts, production and schedules, as well as all the accounting functions, and the machine was barely installed before it was given one of the most important jobs in all its working life. With the aid of that computer — which, it must be said, was gravely mistrusted by many of the older staff who had worked on manual systems all their lives — an extremely rapid rescheduling of production was worked out to boost tractive-unit output and to run down the big rigids. Technically, it was not too difficult, as all the axles were basically common; the gearbox was the same impressive 12-speed unit; brakes were common since they had been standardised after the switch to air systems in the mid-'50s; the

*The S.36 cab was, in effect, a non-tilt version of the S.34. It was introduced in 1966 and was equipped with twin round headlamps instead of the rectangular Cibié type. Some 36s had their lamps in horizontal pairs.*

cabs were basically the same, and so was much of the chassis hardware. There was a problem with engines, however, because at 32 tons the tractive units needed about 30 per cent more power than the eight-wheelers. This led to the wider adoption of Cummins diesels, which were being produced in Britain by that date, and of the new Rolls Royce Eagle diesel, which was being turned out by a new division of the famous company in the old Sentinel wagon-works at Shrewsbury.

The internal change in production emphasis went surprisingly smoothly, and it was only a supply problem on engines at short notice that prevented the operation from being an unqualified success. Even so the order situation did not get too far ahead of production, and only a few sales were lost because the factory could not deliver on time. The engine situation was helped by the development of a further version of Fodens' own two-stroke diesel, the FD6 'Dynamic' engine, which developed 225 hp and was a popular choice in the Twin-Load, and in some tractive units, too. Even a few eight-wheelers had that 225 hp engine, and this was an unprecedented output for a 24-ton machine.

While the UK market demanded ever more complex trucks, with more power, three-line air-brake systems, load-sensing valves, and all the sophisticated hardware which is inseparable from modern trucking, export activity blossomed quite suddenly after years of sustained effort, particularly on the part of Ted Foden and his brother Reg. These export markets demanded much simpler trucks than the British and, despite the competition from American and European truck-builders, in 1965 about 21 per cent of Fodens' total turnover of £7.48 million consisted of export sales. Dumpers were particularly saleable, since many emergent nations were beginning programmes of industrialisation and civil engineering construction. That year a new 35-ton-payload, six-wheeled dumper was introduced, developed from the earlier 24-ton type; this design is still an intrinsic part of the current FC35 series, although more powerful engines are now used and detailed chassis changes have been made.

The secret of the double success at home and overseas was that the mechanical components were all the same and, from the limited number of units turned out by the component and machine shops at Elworth, either home-market or export machines could be assembled at will. The computer control helped, of course, but it was the rationalised engineering which gave the company the necessary flexibility to deal with fluctuations in demand both at home and abroad.

It was during the early 1960s that the South

*The American-based company, Ryder Truck Rental, based its new UK heavy-truck operation on 250 hp Foden tractive units with S.40 steel cabs, when it set up in Britain in 1972. Ryder's chief engineer, Tom Mannix, went on record as saying that Foden was the toughest truck in Britain.*

African subsidiary company, which had been operating at very low turnover and negligible profit for a number of years, began to produce more satisfactory results. New road-building programmes to replace a shrinking rail system brought a need for dumpers, followed by a steady growth in the on-highway truck market. Fodens' patience was rewarded — they were able to meet a rapid increase in demand with ease and, what is more, to back it up with good service support.

New export markets continued to bring in good overseas earnings for the remainder of the 1960s, in particular the rapidly developing oil-producing states of the Middle East, while at home a continued market growth was the product of another short period of political stability. A new version of the stylish fibreglass cab was introduced at the 1966 Show, with twin headlamps instead of the rectangular design and a fixed mounting instead of a tilting arrangement. This was called the S.36 cab, and it gave the operators a choice of mounting, as by no means everybody was convinced of the benefits of tilt cabs at that time. The greater power called for by operators led to an increased use of Cummins and Rolls Royce engines, with outputs commonly up to 220 hp, and occasionally up to 250 or 260 hp for heavy haulage machines. At the same time a reduction in the number of contracts for the two-stroke engines from the Royal Navy, NATO and overseas naval services meant that two-stroke production became very costly, and the number of such engines fitted to trucks steadily declined.

Another version of the fibreglass cab was produced to suit the tipper models. Its main feature was a divided screen, which reduced the cost of the frequent screen breakages experienced in that kind of work. It was also a little taller than the S.36, and was given the code-name S.39. At about the same time two new steel cabs were introduced. One was made by Motor Panels Ltd in Coventry, and with a Foden frontal treatment it still looked like a Foden. But it was made in steel, and was very much stronger than the fibreglass type, especially in accidents at speed. This was the S.40 cab, and a further type called the

**Right** *Some road-going trucks were fitted with the angular S.50 half-cab in the early 1970s, although it was primarily intended for dumpers and tipper chassis.*

**Far right** *In the early 1970s the prospect of higher gross weights led many operators to put five-axle artics into service that could later be up-rated from 32 to 38 or 40 tons. The change never took place. This forward-angled screen cab was the S.60 type but it never became popular.*

*The postwar expansion*

S.60 was offered with a forward-sloping screen, built in steel entirely by Fodens, but its peculiar looks limited its popularity. In addition there was a half-cab version of the S.60, called the S.50, which was quite popular with tipper operators and some short-haul artic users, mainly because it was much cheaper than a fully styled cab, and less prone to damage in that rough-and-ready class of haulage. The chassis engineering underneath all the cabs remained faithful to the rationalised system worked out in the early 1960s, and little change beyond detail improvements took place. However, chassis-type numbers underwent a change, largely because of the computer's complex 'thinking'. Each chassis type had a computer code of impossible complexity for mere mortals to remember, but those were simplified to a series of letters and numbers for everyday use which told a great deal about the vehicle. For example, the Gardner-engined eight-wheeler with a 'translated' designation of RG18/30 was officially labelled the 08R.030.G18.83, with a further suffix after that for detail item changes. The 'human' code was quite simple to crack: in this instance the R meant it was a rigid chassis — an artic would have an A — the 18 meant 180 hp, and G stood for Gardner engine. The number after the stroke gave the chassis gross weight — 30 tons in this example. Similarly an AC29/38 would be a Cummins-engined tractive unit for 38-ton work with a 290 hp engine. The system led to a certain amount of initial confusion among dealers and customers who had for years been accustomed to the old FG6/15 type of model number. However the new system was logical and told more about the vehicle, so it was retained, and is still used today.

A series of events took place in 1970 that could have drastically altered the shape and future of Fodens Ltd. These events involved the extraordinary takeover battle that raged over the small but highly profitable Atkinson Lorries Ltd in the second half of 1970. ERF started it all off on June 17 by making a bid for Atkinson worth about £2.8 million. After an increased offer from ERF, Fodens put in a counter-bid on July 30. Both Fodens' and ERF's bids were rejected, but there followed what amounted to an auction, with the two companies trying to outbid each other. Fodens' final bid was made on September 18 and was worth about £4.2 million. ERF stayed in a little longer, but it was Seddon that finally got the prize, only to be taken over itself by the International Harvester company a year or two later. This extraordinary takeover battle is still talked about today, and many authorities reckon that Fodens or ERF would have made better parents, for at least they would have been much less likely to admit American owners to a large section of Britain's independent truck industry. All manner of conjecture surrounds that takeover. If Fodens had succeeded in buying

Atkinson, would they have spent money on the massive manufacturing facility that was partly responsible for the 1975 cash crisis? We shall never know, of course. The Atkinson affair is more fully discussed in the author's book, *Seddon Atkinson* in the PSL *World Trucks* series.

Fodens Ltd continued to adhere to the policy laid down by William Foden in the 1940s of making as many components as possible under its own roof. But changing economic climates brought about new concepts in manufacturing, and by 1970 most medium-sized truck manufacturers were aiming at no more than 50 per cent self-manufacture content, some considerably less than that, whereas Fodens were still making practically everything except the engine and the electrical equipment. This policy undoubtedly meant tighter quality control — although even so there were occasional lapses, notably with clutches and cabs — but the cost of producing specialised components in the low volumes needed at Elworth raised prices considerably. As a result, by 1972 Fodens were some of the most expensive trucks in Britain with the price tag of around £20,000 for an eight-wheeler nearly £3,000 above the British average. This price margin became greater when the new S.80 cab was introduced in 1972, because the internal strengthening needed to give an acceptable degree of collision-safety made it expensive to produce even in glass fibre, a material geared to low-volume production.

The S.80 cab had a number of good features, but it was not one of Fodens' most popular designs. Access was not as good as in the latest European cabs, and the hub-ring on the wheel still had to be used as a step. It tilted for access, but was locked down by two screw-pins under the back of the cab, and replacing these made the mechanic or driver's clothes filthy. On the credit side the cab was spacious and comfortable, it had an excellent instrument panel, an adjustable steering-column and a swing radiator behind the hinged grille panel, so that most maintenance jobs could be done without tilting the cab. By then the two-stroke engines had been discontinued for vehicles, and were retained mainly for marine purposes, so the engines in the S.80 types were all Cummins, Rolls Royce or Gardner. Two new gearboxes were developed from the 12-speed type, and used the same principle of a four-speed main box with an epicyclic auxiliary on the back; in fact a lot of the parts were common. The eight-speed box was essentially an on-highway unit, with a bottom gear of 7.25 for normal use, and an emergency crawler at 12.25 to 1, extending up to an overdrive top gear of 0.77 to 1. The 9-speed unit was mainly for specialist work, where lower axle-ratios would be used, and had a gear-ratio range from 8.05 to 0.77. These boxes were much easier to drive than were the old 12-speed as they had logical shift patterns, although they were regarded as 'kids' stuff' by long-term 12-speed men who could play tunes on the old difficult-to-handle boxes. The rationalised axles were continued, which helped to keep costs down.

But new moves were afoot. Late in 1969 negotiations were completed for a large piece of land on the opposite side of the main railway line from the works, and planning permission was eventually obtained for new production plant on the site. It had been decided that the only way to compete economically in a toughening market was to install the most modern manufacturing plant available, and step up production at a lower unit-cost. No doubt Edwin Foden would have shuddered at the idea, but these were the 1970s, not the 1870s, and it was the only way to survive.

*An elaborate version of the two-stroke Foden diesel was the FD12 Mk VII — turbocharged as well as mechanically supercharged — used by military authorities in assault craft.*

# Chapter 17

# Expansion, crisis, recovery

As the 1960s drew to a close, there was a steady increase in the volume of British truck sales both at home and in the traditional export markets — the major Commonwealth countries as well as various emergent African states. There was also a rising demand in the oil-rich Middle East. The increased business was enjoyed not only by Fodens, but also by the other manufacturers like Atkinson, Seddon, ERF and Leyland.

Despite the fact that the premises at Elworth had been expanded several times since the war, and now stood at a total of about 50 acres, there were unacceptable restrictions on production. Much of this area was occupied by component and sub-assembly manufacturing facilities, since Fodens was one of the few truck-builders to use as large a proportion as possible of its own manufactured parts. These included the gearbox, clutch, axles, cab, most of the chassis hardware except the springs, and even detail work like upholstery and so on.

So, despite the considerable size of the complex, maximum manufacturing capacity was only about 40 chassis per week at the outside. If this figure seems rather low in relation to the size of the plant, two factors should be remembered. One, of course, is the high self-content of the Foden vehicle. The other is that a high proportion of Fodens' output was, and still is, taken up by large, complex, multi-axle chassis and these occupied much more time and space in the production process than did the simpler four-wheelers and tractive units turned out by the majority of Fodens' competitors in Britain.

Fodens were first off the mark by a long way in building chassis to suit the revised Construction and Use regulations which came into force in Britain in 1972. These made the rigid eight-wheeler a much more attractive proposition economically than it had been, by increasing the permitted gross weight from 24 to 30 tons. Fodens needed very closely controlled chassis designs in respect of axle-spread and bogie-spread dimensions, and weight distribution was fairly critical too. But they had suitable chassis available in time for the Scottish Motor Show at Kelvin Hall in November 1971, even though the regulations were not publicly announced until later, and did not come into force until June 1972. The effect of this change in C & U, and of the buoyant economic situation in Britain at the time, was a big increase in orders, not only for the expensive rigid eight-wheeler trucks, but also for the four-wheeled tractive units. The latter enjoyed relaxed dimensional controls under the new regulations, which meant they could now operate with four axles at 32 tons over a wide range of dimensions, whereas previously five axles had frequently been necessary.

The Fodens board had been studying the forward market and their own production capacity: it was quite clear that within a couple of years the company would be unable to sustain the desirable growth rate made possible by improved facilities. So they decided to embark upon a major production expansion scheme which would give Fodens adequate capacity for foreseeable chassis demands well into the 1980s or even the '90s. This was to be no ordinary reshuffle to find a bit more room here, an extra corner there, as many of the previous reorganisations had been. An entire new assembly complex was to be built, with the most modern processing and unit-storage facilities, and ideal working conditions. The planned capacity was to be at least 80 chassis a week on single-shift working, or considerably more, if required, on multiple-shift work. The space vacated in the original works by the old chassis assembly area would be used for extra machining capacity of mechanical components, to feed the main assembly plant.

Not only would the new plant have to step up production, it would need to be extremely flexible, capable of building anything from a 100-ton multi-axle export dump-truck to a home market 20-tonner, and everything else in between on two, three or four axles, with from 150 to 450 hp, in any combination. A sophisticated degree of control was therefore a major factor; in fact, it was a job for a computer. This meant it was going to be expensive.

The actual design of the installation was quite difficult to work out. Very few truck assembly plants as large and as sophisticated existed in Europe. American methods were not, on the whole, suitable in this case, and accurate cost estimates were difficult to pin down. Various plants were examined in the search for the right answer. The then-new Daimler Benz truck plant at Würth was impressive but its capacity of about 140,000 trucks a year was wrong. The plants at the various Ford and Bedford factories were not then geared to super-heavy chassis, and most of the others were too old-fashioned to teach much in the way of advanced truck production — a surprising number still had systems where chassis were pushed by hand along an assembly track on little bogies, as had been general practice in the '30s.

But there was one plant which would repay careful study, and that was in Sweden. Some years earlier the Scania-Vabis company had

**Left** *The Foden works in the early 1970s, before the big expansion: Elworth road can be seen running right through the middle, and the line of the old North Staffordshire railway at the left should be compared with the drawing of the works in Chapter 4. The service department is nearest the camera with the 1956 office block and the playing fields at top centre. The new assembly plant was later built off-camera to the left on the other side of the main railway line.*

**Above right** *Old and new factories are of about equal size. The new plant was built in the early '70s beyond the railway line, with a test track to the south of it.*

*Expansion, crisis, recovery*

realised that production demand would soon outstrip the available manpower in their area, so they built a new plant which would give them increased production with a modest workforce. Much of the basic concept was their own, as Scania had (and still have) a very elaborate technology and methods department which developed all manner of production-control and management systems for internal and external use. One of the interesting aspects of the Scania-Vabis project was that much of the actual hardware and control system was designed and built in Britain.

The plant consisted of a system of overhead conveyors, fed by 'sidings' from which major components and sub-assemblies emerged, so that a heavy truck chassis could be built up steadily, efficiently and easily from a bare frame up to the completed machine. Until the final stage the chassis remained suspended at various heights above the shop floor, each variation in height being dictated by the work to be done at any particular time.

This system of heavy truck production seemed ideal to the Foden engineers. Not only did it provide great flexibility in the type and sequence of chassis to be built, but also the basic installation was relatively simple and inexpensive compared with some ground-level conveyor systems. There was another great advantage in that, if a change of layout, or further expansion, was needed at any future date, it could be accommodated with little disruption of the basic facility. By the middle of 1972 all the plans had been discussed and drawn up, all the costings analysed, and all steps of the project double-checked to make sure that the new plant would do what was needed. Everything tallied perfectly, and the final decision was made to go ahead.

There was little time to waste, for already there were signs that the inflationary spiral of the mid-'70s was beginning; although in the autumn of 1972, with the boom in truck buying, there were very few clues as to how quickly prices would rise within the next 12 months. The estimated cost of the expansion was £3.8 million and, if the project were postponed for some years, who knew what the inflated cost might then be?

Just how well the plans had been worked out was illustrated by the fact that nine months after building began (on a site alongside the existing works on the other side of the railway), the first chassis rolled down the ramp at the end of the line and was driven round to the inspection and detail-adjustments bay.

As an industrial-design project the new Fodens' plant was a triumph. It ran like clockwork from the very outset with very few problems, and the working conditions were far superior to anything that had been seen at Fodens before. There were, however, some severe supply problems. One outstanding feature of the plant was that it combined the most up-to-date concepts in process machinery and production control, with the best of the traditional methods of hand-assembly of trucks, which Fodens had developed over a period of 70 years.

The new shop consisted of a continuous overhead conveyor, with a surface track extension at the 'finished-chassis' end of the loop, and various storage and feed facilities flanking it along one side. Finished chassis frames, erected on the floor of the frame shop in the same building, were picked up by the overhead conveyor and the embryo vehicle began its assembly journey. The next two stations saw rear and front axles moving in on 'siding' overhead conveyors, suspended at exactly the right height in relation to the frame, and the fitters working at that station attached the axles to the frame. In a sub-assembly lane, operating on an identical basis, engines were assembled to clutches and gearboxes before entering the main-stream conveyor, where they dipped down to chassis level for installation in the growing vehicle structure. The innumerable items of chassis hardware, such

*Expansion, crisis, recovery*

*Various stages of chassis assembly in the new plant at Elworth.* **Left** *Frames and axles are lined up in a side bay at the beginning of the process.* **Top left** *Axles are mounted to the chassis as it hangs from the slowly moving overhead conveyor.* **Centre** *The engine comes in on a side conveyor to be lowered into the chassis.* **Below left** *The chassis turns on to the second leg of the conveyor as items of chassis hardware are added.* **Above right** *The near-complete chassis is painted.* **Below right** *Finally on the floor, the chassis receives its cab.*

*Anxious hands reach for the very first S.90 steel cab in the experimental shop, as it is lowered on to the prototype Universal chassis. The date was 1974.*

as auxiliary power take-offs, special piping and filter equipment, suspension restraints, brake valve gear, tipper subframes, fuel tanks, and so on, gradually transformed the skeletal truck into something resembling a finished machine, and, when these had been fitted, each chassis was prepared for painting and then passed through a painting and baking lane. On emerging from this series of booths and ovens, the chassis were complete except for cabs and controls, and they finally touched down on to firm ground after having the appropriate wheels and tyres attached. The final length of ground-level conveyor saw the cabs fed in from yet another 'siding' and lowered on to the chassis. The numerous controls were installed and adjusted, followed by the final electrical circuitry which would bring the vehicles to life. That last floor conveyor is 230 ft long, relatively short in relation to the 925 ft of main overhead conveyor and almost as much again in sidings. The standard chassis production from the very first stage to a completed running vehicle is 14 hours. The main installation was designed and built by MHS International Ltd, a British firm from Watford, Herts, who had also done most of the work on the Scania plant. Control of the step-by-step production processes, and the marshalling of components in the correct order for incorporation in the chassis is looked after by an IBM 370 controlled storage system.

So the first few chassis came rolling off the line in the late summer of 1973. For a time production lagged well behind demand, as new tooling in the component shops, a new cab, and improved gearbox and clutch designs were added to the overall reshaping of the Foden truck range. By the end of the year, however, all was going well: the new production facility was settling down and a steady rate of up to 50-55 chassis a week was programmed.

The total investment, including tooling and improved components, was more than £5 million, but it was worth every penny. Fodens now boasted one of the most modern truck-production plants in Europe, even though it was far from being the largest. The shareholders were happy at the prospect of an expanded company battling for the truck market in an increasingly competitive world. In Britain, this market had been heavily infiltrated by imports from firms such as Volvo, Scania, Daf, and Mercedes, soon to be followed by Magirus, MAN, Fiat and Saviem. It looked as though a real fight was on the cards, and so it turned out. However it was not to be the marketplace fight that everyone had been looking forward to, but a fight for the very life of the old established firm.

The winter of 1973-74 saw Fodens stepping up output as never before, making good use of their new facility. The fleet re-equipment programmes which had followed the change in Construction and Use regulations in mid-1972 were continuing; the order book was still healthily full. But there was an uneasiness in the transport industry, indeed in the economy as a whole. Industrial disputes in key industries such as coal-mining and electricity had seriously affected the national economy and had led to a great deal of short-time working. A series of dock strikes had played havoc with the balance of payments as export goods piled up waiting for someone to load them aboard idle ships. Only the roll-on roll-off ferry traffic kept commerce going. It was a bleak winter on the industrial front, although fortunately the weather proved uncommonly mild, making the mining and power industry strikes less painful than they might have been.

But events took their toll: by the time the first green leaves of spring emerged hesitantly across the Cheshire plain, the mood of optimism that had pervaded every corner of the Elworth factory in the autumn had vanished. The truck market had virtually collapsed. Those truck orders that had been outstanding vanished almost overnight. Many were cancelled altogether as transport firms decided to tighten their belts and make do with the vehicles they already had, instead of buying new or additional ones. Other orders were snapped up by hungry importers, desperate to sell chassis and not greatly caring whether or not they made a profit, as in most cases their UK operations were subsidised by much larger ones back home. The alarm bells rang early at Elworth, and by March 1974 a cut-back in production and forward ordering of parts and materials was initiated. But the bells had not rung early enough.

Fodens' practice of making a high proportion of their own components, meant that forward-ordering lead times were very much longer than those of most of their competitors. It was not simply a case of ordering, say, 100 gearboxes or 200 axles, and then cutting back on those orders a little later if necessary. At Fodens the lines of supply were very much longer. Raw materials for the foundry, stock steel for the machine shops, bulk polyester supplies for cab construction — these would all be ordered and on the premises long before a manufacturer using outside parts would have ordered his supplies. Consequently, as the summer sun steadily warmed the English country-side, the climate at Elworth became rapidly colder and gloomier. The market had taken an abrupt nosedive and, owing to the Foden method of making trucks, it was not possible to reduce the rate of production at a corresponding rate.

The company had also embarked on the development of a new heavy-truck range — called at that time the 'Universal' — aimed at export markets both in Europe and elsewhere. Although Fodens' development programmes, even then, were by no means as elaborate and costly as some of their competitors, there was undoubtedly a lot of money involved, probably something in the order of £750,000. The inevitable results could be seen coming a long way off. Fodens were heading for a cash crisis, and heading for it fast. Stocks of chassis grew alarmingly in the yard at the back of the plant. The dealers and distributors took as many as they could and paid as much as they were able, but they too, like the rest of the industry, were struggling with cash-flow and liquidity problems.

The ship finally hit the rocks just before Christmas 1974. The trading position had been poor, although no poorer than anyone else's, but the loan repayments for the production investment programme had to be made out of reserves instead of from increased production revenue. There was no way of reducing the bank overdraft as had been planned. The company's launch into Europe in conjunction with the German firm of Faun had foundered, since local legislation brought the construction industry in central Europe to an almost complete standstill. Fodens' production method meant that they employed nearly 3,000 people, whereas other factories who built from bought-in parts employed about half that number to produce the same volume of finished vehicles. The wages bill was consequently very high. William Foden went to the company's bankers for further help to tide them over the rough patch. The bank politely but firmly said, 'No'.

Somehow the story reached the press, appa-

*First production Universal chassis were heavy tipper versions, fitted with 335 hp Cummins engines with Foden eight-speed gearboxes and hub-reduction axles, and were introduced in 1974 at the Amsterdam show.*

rently through local gossip picked up by a reporter on a local newspaper, the *Sandbach Chronicle*. The entire industry was aware that Fodens Ltd had hit cash problems, and was in danger of winding up. But, as it was Christmas, none of the financial or government offices was open, and the board of directors, the shareholders, and by no means least any employees who had not already been made redundant, spent an agonising fortnight waiting to see what would happen. Just what could be done? Was there any way of saving the great company which had been the centre of industrial and social activity in that part of Cheshire for well over 100 years? Nobody really knew, but one thing is quite certain: there was precious little Christmas cheer to accompany the roast turkeys and plum puddings in Elworth and Sandbach that Christmas.

As soon as work resumed after the festive season Fodens' management team went into action. Their first port of call was the Department of Industry, to discuss the company's problems with government and treasury officials. The argument that they offered was based on Fodens' good long-term orders, including several military and export contracts spread over two years or more. The team explained that they had a very modern production plant, a great deal of stock both of complete vehicles and components, and a new model range on the stocks which showed considerable promise. They also believed

*A limited number of special tractive units, capable of operating at up to 100 tons gcw, were built in 1974-75, powered by Rolls Royce Eagle 290 engines, and using tandem hub-reduction axles on the bogie. This one operated with a King low-loader. There was also a 60-ton version.*

*Expansion, crisis, recovery*

that they still had the confidence of operators both at home and overseas — in fact all they lacked was ready cash to pay the weekly wage packets. In this last, of course, they were by no means alone.

Many criticisms had been heaped on the government of the day, a Labour government led by Mr Harold Wilson, but it had set up some very constructive industrial policies. One of these was a contingency fund aimed at helping out firms which found themselves in difficulties, particularly as a result of investment programmes. This fund later became known as 'lame duck lolly', and was undoubtedly misused in a number of instances. But in those first days of 1975 it was like a lifeboat to the ailing Foden concern, and a loan was agreed, providing that a government investigation of the company showed a basically sound future.

The first examination took only a few days and Fodens were awarded their government loan. The precise figure was never actually made public, although it is reasonable to suppose that it was something between £1 million and £1.2 million, certainly no more. But it was enough to get things going again, pay the workforce and the heating bills, continue with new-model development and, above all, to continue trading. People breathed again, even though they all knew it might only be a stay of execution. Meanwhile the Department of Industry engaged the accounting firm of Price Waterhouse to conduct a detailed examination of Fodens' affairs and find out whether the company was a suitable subject for long-term government support. For weeks the accountants dug and probed into every corner of Fodens' affairs, not forgetting the overseas companies in South Africa and Australasia.

In the end their report was favourable, but their help was not needed. For in the meantime the Foden board had been busy. They believed that with a basically sound company like theirs they ought to be able to raise some money in that admirable financial institution known as 'The City'. The bankers had advised that an injection of at least £3 million of new capital was required and, although the money market was very tight at that time, an attempt was made to raise this sum. For weeks the financial press columns had been talking about Fodens, and opinion was divided about 50/50 on their chances of success. There were, after all, numerous other companies of equal standing that were in financial difficulties. Also, Fodens' last full year's trading results before the crisis had shown a pre-tax profit of a mere £230,921 from a turnover of £22.6 millions, and that was not going to excite many investors.

Nevertheless, such was the support given in the City to a proposed Rights Issue of £3.17 million cumulative preference shares (convertible to a new class of ordinary share, should the shareholders choose), that by the spring of 1975 the financial rescue was accomplished; the rights issue was completely taken up, helped in the latter stages by the publication of the Price

**Above** *Spring brakes were included in heavy-truck specifications from 1968 onwards.* **Top right** *Helically ribbed brake drums are a Foden feature, aimed at promoting airflow across the drums.* **Bottom right** *Hub reduction axles up to 20-ton capacity are used on heavier chassis including dumpers.*

Waterhouse report which, in effect, gave Fodens a long-term clean bill of health.

While Department of Industry intervention would have been acceptable as a last resort, it would effectively have meant nationalisation, an almost unacceptable solution to a family institution like Fodens Ltd. This prejudice undoubtedly lent an added impetus to the management's fund-raising efforts in the City, and the steadfast confidence of the shareholders in William and David Foden and their fellow directors carried them through the storm and on into calmer trading conditions.

Despite the good news on the financial front the factory was far from in the clear. The gap between decreased demand and increased production was now too wide to be closed quickly. Throughout 1975 and early 1976 the young board strove to improve the situation, seeking new export business for the heavier specialised trucks and developing the military side, although some of the home-market models had by then lost much of their appeal in the face of newer designs from other manufacturers. Consequently, despite encouraging long-term orders and a promising new model range in development, the immediate trading position was far from satisfactory. Considering the problems which the company

*Expansion, crisis, recovery*

had faced and was still facing a trading loss for 1975-76 was inevitable.

The results were worse than most authorities had forecast, and there was a flutter of excitement and more predictions of gloom and disaster in the press when the balance sheet showed a pre-tax loss of £1,010,886. The new chairman, Leslie J. Tolley, was at pains to explain that there had in fact been a net trading profit of £357,800 — down from £1,950,722 the year before — but the heavy loan interest and repayments had produced the loss figure. Turnover was fractionally up at £28.6 million and the message was that the crisis had 'bottomed out'. Not everyone was convinced but Fodens fought on, confident that the worst was over.

The first really good news for a long time came in April 1977 when the results of the year's work showed a pre-tax profit of £1,738,015, with a trading profit of £3,139,300 from a £47-million turnover. At the same time orders for military vehicles worth over £10 million, and construction vehicles worth more than £2 million were announced, and the City speculators' faith in the name of Foden was seen to be justified.

Fodens were not yet out of the woods. Home-market trading continued to be depressed, and the steady return to 1973 levels of UK truck sales did not really begin until late in 1977. A factor in this delay was that the full resources of the plant could not be utilised because of supply problems. One of these problems was with the steel cab for the newer models — the S.90 and S.93 type cabs. These were to be supplied by the Motor Panels group, gradually replacing the Foden-built, glassfibre reinforced polyester cabs of the S.80/83 type. Even four years after the nominal introduction of the steel types, supplies were still far from what was needed, and many military vehicles were delivered in 1976-77 with plastic cabs instead of the specified steel ones, on the understanding that they would be updated later. A switch in manufacturing policy as partial insurance against another cash crisis was undertaken in 1976-7. The new policy was to employ proprietary components, such as the Eaton-Fuller gearbox and Rockwell axle, on some models, so cutting down the long lead times and consequent cash-sensitivity of production. This proved to be something of a two-edged weapon when labour disputes in the factories involved held up production yet again, but, despite all these problems, Foden was back to reasonably healthy profits by 1977. The first-half figures for the

**Left** *A prominent feature of the S.80/83-series models was the radiator, which swung forwards to provide access to the front of the engine. The cab could be tilted for major attention.*

**Right** *Foden gearboxes in 12-speed, nine-speed and eight-speed versions all use a main four-speed box with an epicyclic splitter. The manual shift mechanism is at front right with the air cylinders for the splitter shift behind it.*

1977-78 trading year showed a pre-tax profit margin of £1.28 million to October 1977. A forecast for the whole year of £2.5 million was made at the time of the Rolls Royce takeover bid, and in the event that forecast was exceeded by a comfortable margin. The recovery, so earnestly prayed for in that dark winter of 1974-75, was finally a fact.

At this time a new range of very attractive and efficient highway trucks began to appear in large numbers on the highways of Europe — the Haulmaster and Fleetmaster models. Bill Foden had forecast just such a turn of events in an exclusive interview with the author for the monthly journal *Truck*, in October 1975, when he said, 'Our development programmes have been set back by about 15 months by the crisis, but that will at least enable them to have had the longest and most exhaustive testing and development that a Foden has ever had. But by 1977 they will be thoroughly developed and ready for the outside world, and by then we believe that hauliers will once again be buying trucks, and we'll be there ready for them.'

Fodens were indeed ready, and the new models had an extremely favourable reception when they were launched in the autumn of 1977. Just as Bill Foden had forecast, this was the time when the market was picking up after what amounted to a four-year recession.

Throughout the months of crisis there had been fears that, even if the old company were to pull itself up once more by its bootstraps, and return to a safe trading position, it might be swallowed up in a takeover deal, as had many other old firms. The early and mid-'70s were prime years for the destructive process of take-over, asset-stripping, and regurgitation on the part of big multi-national groups and a number of property companies. It was suggested at one stage, and the idea received a certain amount of government support, that Fodens at Elworth and ERF at Sandbach should merge to form one company, but neither partner was willing. In any case it would have been a difficult manoeuvre, as ERF are basically builders of trucks from bought-in components, unlike Fodens.

Early in 1977 Rolls Royce Motors made a series of bids for Fodens, but such was the fast-returning strength of the firm, and the loyalty of the shareholders (many of whose families have held shares since Fodens Ltd was formed in 1902) that the bids failed. The shareholders were adamant that no Johnny-come-lately outfit like Rolls Royce, which had itself foundered over its aero-engine business and been divided up and reformed in 1971, was going to take them over. Rolls Royce were also at that time attempting to acquire as much share capital as possible of the Gardner engine company, in Manchester, but although they succeeded in getting hold of about 16 per cent, the Hawker-Siddeley group eventually acquired Gardner, and the Rolls Royce plans for a sizeable truck-manufacturing empire finally foundered.

By the late 1970s the take-over mania had faded into a minority activity. But perhaps the major security that Fodens have against further financial difficulties, take-overs and erosion of their competitiveness, is that their plant, still one of the best in Europe, was installed at 1972 prices, while the majority of their competitors are having to re-equip and update at today's inflated prices. This fact is a foundation-stone on which the secure trading of Fodens Ltd should be supported for many a long year to come.

*The Foden 'Sixer' chassis for construction work broke new ground by using a high proportion of proprietary parts. It was introduced in 1976 and weighed comfortably under six tons. Most Sixers work as mixer chassis, though a few have been built as tippers. The old S.39 cab was employed.*

# Chapter 18

# Into a new age

There were many who had forecast the end of the ancient and traditional firm of Fodens Ltd at the time of their financial crisis in 1975-76. Not least among these were some of the daily newspapers which, stirred by the thirst for sensationalism, ran headlines on their business pages to the effect that foreign truck imports had killed Fodens, or that labour disputes both inside Fodens and at their suppliers had brought about the demise of Britain's oldest commercial-vehicle manufacturer. In the event, none of them was right. Fodens did not die; although, at a dinner in London in 1977, David Foden admitted that a caption carried by the monthly journal *Truck* to an article about Fodens, '. . . nearly died, better now thank you,' was nearer to the truth than any of them cared to admit at the time. Following the injection of new capital the new board, headed by L. J. Tolley with Bill Foden as chief executive, and aided by David Foden, Patrick Twemlow and a loyal, determined workforce, saw the company back on to an even keel, and the trading figures rose remarkably rapidly to what a company turning over around £30 to £40 million a year should expect. With this recovery came a change in basic policy that makes a great deal of sense in a modern industrial climate — to reduce the content of self-manufactured components by drawing on outside specialist suppliers for the more complex items.

The first truck chassis to be built under this policy was the unusually named 'Sixer', which appeared in prototype form at the 1976 Earls Court show. At first there was considerable ribaldry over the 'Sixer', as it used the old S.39 fibreglass cab which had gone out of production many years earlier. There were good reasons for this, however, one being that the big square S.80/83 cab was too bulky for tipper and mixer work, and the old design did the job admirably. But there were other changes. Eaton axles were used for the first time, and Girling brakes appeared too. Fodens had been drawing engines from outside suppliers for many years of course, but numerous smaller items such as springs and brackets were all bought in. The result was a remarkably effective and lightweight chassis, easily capable of carrying a 16-ton load of wet concrete or 17 tons in a tipper body, with all the spectacular economy of the Gardner 6LXB engine to back up the sound economics. But impressive though this model was, it needed much more to re-establish Fodens Ltd on the path to strength.

The military contracts obtained during 1974 and '75 provided splendid bread and butter at a time when confidence among UK truck users was low. In fact, the early Foden military trucks were so successful that several more contracts followed, totalling more than 2,000 trucks to be delivered over a period of about five years. The contracts were for six- and eight-wheeled general-service and medium-mobility trucks, plus a limited number of 4 × 4 trucks with a very high cross-country performance ability. Although delays did occur in fulfilling these contracts, mainly through erratic supplies of components, particularly cabs, the in-service record of the trucks has been excellent, and further on-going contracts resulted. This success was extremely useful to a factory determined to fight back from a weak financial position, and the steady flow of work enabled stability to be built into production schedules for a long time ahead.

Apart from these military contracts, development work continued on a new model range which had been initiated before the crisis. Originally the range was called, rather unenterprisingly, the Universal. It featured heavy-duty chassis for construction work on two, three or four axles, a new all-steel tilt cab and a selection of high-power engines. The Universal range was to extend the newly formed policy of employing some proprietary components instead of self-

**Left** *David Foden hands over the keys of the first Foden eight-wheeler to be delivered under the military contracts to Brigadier H. R. Dray in July 1975. The Foden band played at the ceremony, held near the site of the War Office Trials which had taken place more than 70 years earlier.*

**Below** *The prototype 4 × 4 military truck travelling at high speed in wet and difficult conditions. The stability and handling of this truck is outstanding, and it reaches speeds in excess of 60 mph.*

## Into a new age

manufactured units. The reasons for this policy were rooted in the complex matter of labour intensity. Obviously, if a company makes everything itself down to the last nut and bolt, it can control the quality and the quantity, but the necessary investment places too great a burden on a small to medium-sized firm. At the other end of the spectrum, a company which relies entirely on outside supplies is too vulnerable to market changes, and has no influence on forward development and new designs. Somewhere in between is an ideal compromise, and the precise position of that compromise varies according to the size of the company. Certainly with a company like Fodens, employing about 3,000 people to produce something like 2,000 chassis a year, the labour intensity was too high. The manufacture of brakes, gearboxes, cabs, axles, in fact everything except the engine and detail accessories like air cleaners and electrical parts, was a luxury finally ruled out by modern economics.

As the modern development programme based on the Universal progressed the heavy machines orientated themselves towards export markets, and a new derivative range emerged for the home market and Europe. The larger, more powerful model is the Fleetmaster; a 38/40-tonne tractive unit with a 290 hp engine from either Rolls Royce or Cummins, a Fuller Roadranger gearbox and a Rockwell single-reduction axle, Lipe Rollway clutch and Rockwell two leading shoe brakes. The development programme was reshuffled due to the cash crisis but in the end this proved beneficial. When the Fleetmaster finally emerged in the autumn of 1977 its impact was immediate. The prolonged development period meant that all the bugs had been eliminated from its system, or at any rate more than can usually be eliminated by the early production stage. After detailed test programmes the technical press were quite excited about the Fleetmaster. Typical comments were, 'The ride was one of the finest we've come across with an artic, and that was achieved without

*A 6 × 6 Foden medium-mobility truck demonstrates its cross-country ability fully loaded and drawing a trailer. Power is provided by a Rolls Royce Eagle Mk III 305 hp diesel.*

*In the early summer of 1977, prototype Fleetmaster models began to appear on the roads of Cheshire on trials with part of the front panelling removed as an attempt at disguise. But clearly a new model launch was imminent.*

*Foden re-entered the bus business in 1976 with a new design powered by a rear Gardner engine. The chassis used a perimeter frame, stiffened by the body structure, which was built by Northern Counties Coachworks in Wigan.*

suspension seats', and 'Control was totally positive and accurate, even storming up and down long motorway hills at flat-out speeds in the pouring rain.'

High performance with good economy, and above all an astonishingly simple chassis layout, attracted the operators, and immediately a steady stream of orders began to flow into the works. Production deliveries began early in 1978 and from all accounts of those trucks in service, the promise of the prototype chassis is borne out on the road. The Fleetmaster has a smaller brother called the Haulmaster, which is a hybrid between the old Foden 'we-make-it-all' models and the new buying-in concept. The engine, cab, suspension and steering, together with a great deal of detail hardware, are bought from outside, but the axles, clutch and gearbox are still Foden-built. The aim is to use proprietary parts in tractors and continue with all Foden manufacture of the heavier types. Haulmaster models use lower-powered engines, mainly the 255 hp Gardner and 250 hp Cummins units, and externally they are distinguished by a divided flat-glass windscreen in place of the one-piece curved screen of the Fleetmaster. The Haulmaster is a much more traditional type of truck than the Fleetmaster, and undoubtedly aims to attract the many British truck operators who view anything remotely

advanced or progressive with the gravest suspicion.

The Haulmaster subsequently evolved into multi-axled heavy chassis for the home market with the Fleetmaster carrying tractor colours, while the export mainstay was the impressive six-

*Into a new age*

wheeled Super Haulmaster tractive unit (intended mainly for heavy-duty work in export markets). These models spearheaded Fodens' advance into the 1980s and beyond. Now that the impressive production facility is working on the scale for which it was intended, something like 60 chassis a week are coming off the line. A slow but steady increase in this number is apparent as the markets strengthen both in Britain and Europe. In 1977 a new marketing company was formed in West Germany, called Foden GmbH, serving Germany and surrounding EEC markets, and aimed mainly at selling heavy construction-industry trucks such as the Super Haulmaster. It so happens that the European construction industry is very quiet in the late '70s, due to crippling legislative measures that have all but stopped public building. However, this situation looks like reverting to normal by the end of the decade, by which time the Foden machine should be a good competitor with an extremely attractive price tag.

Back in Britain, the company has re-entered the bus market after an absence of about 15 years, with a rear-engined design embodying a number of unusual features. The chassis has a perimeter frame; in other words, there is a strong steel rail around the outer extremity. It passes over the wheelarches and forms a major structural member. There are floor members, too, but the perimeter frame combines with the body side-frames to form a very stiff box structure. This design was developed in conjunction with Northern Counties Coachworks at Wigan, and the prototype buses are undergoing

*The comfort and driving environment in the Fleetmaster S.93 cab are a far cry from the conditions in steamer days. Compare this picture with the one in Chapter 8.*

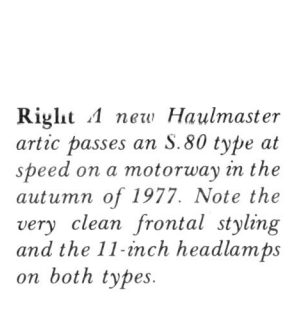

**Left** *Production Fleetmaster tractive units began to appear very late in 1977, and by the spring of '78 a rapidly increasing flow emerged from Elworth with the smart S.93 steel cab.*

**Right** *A new Haulmaster artic passes an S.80 type at speed on a motorway in the autumn of 1977. Note the very clean frontal styling and the 11-inch headlamps on both types.*

*A special drop-centre axle was developed for the new bus chassis in 1976. The driving flange is on the rear of the axle to mate with the rear-mounted engine and transmission.*

extended service trials with the Greater Manchester Passenger Transport Executive. Initial reactions are good, especially where economy and reliability are concerned, although some development will be required to reduce the weight slightly.

The double-deck bus market in Britain is never very large, seldom exceeding 1,500 buses a year and with Leyland, Dennis, Ailsa, Metrobus and Fodens all in the field, some with more than one model, no one manufacturer is going to make a huge fortune. But with each bus selling at nearly £40,000, even relatively low-volume sales can be lucrative.

Further afield, there is still a strong market for heavy trucks in emerging industrial nations such as the oil-producing Middle East states and certain African countries. One of the functions of the German company will be to feed the Middle East markets, since Arab buyers tend to regard Germany as a kind of world truck centre. Indeed, they treat exhibitions like the Frankfurt International Motor Show, the Munich 'Bauma' construction exhibition, and the Hanover Trade Fair with its large transport section, as supermarkets where they select the goods they need for their industrial expansion programmes. In this context Foden GmbH has probably more potential than in purely European sales. Certainly the Middle Eastern markets are vital to any heavy-truck manufacturer and, in addition to the German shopping expeditions by Arab industrialists, there is a constant on-the-spot sales effort in the Middle East itself. In the meantime, the traditional markets like Australia, South Africa and New Zealand continue to contribute significantly to total sales. During 1977 Foden trucks were sold in more than 50 countries, some in small numbers admittedly, but this still represents a wide spread of activity for a small, independent truck-building company.

There are those, particularly in sophisticated industrial communities such as West Germany, or the United States of America, who look at Foden and wonder how it can survive. Small outfits survive, they argue, because of a specialised service facility, while large concerns survive through sheer weight and efficiency. But the medium-size company cannot last, they argue, because it has neither the advantages of personal attention nor the economics of scale. Despite these theories, which are very hard to counter by conventional economies arguments, Fodens Ltd is thriving. It survived in 1933, when a change in the very nature of the transport industry threatened to submerge it. It survived again in 1975, when a combination of economic tricks threatened it, despite the company's underlying profitability and strength. Fodens Ltd did not merely survive, it recovered in a remarkable fashion, with a £2.7 million turnround in profits from 1976 to 1977, and it now has one of the most attractive and saleable ranges of products on the market. There has never been a British commercial vehicle industry without the name of Foden. Who can say whether, in 120 years time, there will still be a Foden in that industry? One thing is certain. Wherever and whenever transport men get together to talk about trucks or wagons or lorries (whatever their particular dialect calls them), Fodens will always be mentioned with a special blend of affection and respect. There is precious little room in transport for legend or romance — it is a hard, tough business at the best of times — but Foden is closer to being a legend than any other name in the business; and so it should be, after more than 120 years' prominence in the transport arena.

# Appendix

# Principal members of the Foden family

Readers may find the following details helpful in distinguishing between the numerous members of the Foden family who were involved with the company over the years. There were, of course, others besides those mentioned here, including numerous daughters.

**Edwin Foden** Son of a village shoemaker, born in Smallwood, Cheshire, in 1841. A staunch Methodist, a great engineer, and a lover of music, he founded the company of Fodens Ltd after taking over the Elworth Foundry from his former employer, George Hancock. Edwin Foden died in 1911.

**William Foden** Eldest son of Edwin Foden, born 1868. He saw the company grow from its agricultural machinery days to become a world famous truck manufacturer, and he was responsible for much of that growth. He was still Governing Director of the company when he died in 1964 at the age of 96.

**Edwin Richard Foden** Second son of Edwin Foden, born 1870. An accomplished engineer and a lover of motor cars. After working with Fodens Ltd for almost 50 years, in 1933 he founded the firm of ERF Ltd with his son Dennis (1900-1960). He was always known as 'E.R.' He died in 1950.

**Samuel Poole Twemlow** Company secretary of the old Elworth Foundry from 1887 and director of Fodens Ltd from 1907; born in 1866 and died in 1928. He married Edwin's daughter Fanny, so establishing a century-long link between the Foden and Twemlow families.

**James Edwin Foden** Son of William Foden, born 1901, and always known as 'Ted'. Except for a brief period in Australia in 1924-30, he was associated with the company all his life, and was a director for 35 years specialising in sales and export work. Joint managing director 1950-71.

**Reginald Gordon Foden** Eldest son of William Foden, born 1899, and associated with the company all his working life. Joint managing director from 1951 to 1971 when he retired. Specialist in managerial, industrial and purchasing affairs. Was always known as 'Mr Reg'.

**Edwin Twemlow** Son of S. P. Twemlow, born 1906. A brilliant engineer, unlike his father who was an accountant. For many years he was responsible for truck engineering at Elworth, and he was joint managing director from 1951 to 1972, when he retired.

**William Louis Foden** Son of R. G. Foden, born in 1929, and a specialist in accountancy and financial matters. He was made an executive director in 1969, full director in 1970, chairman in 1973 and chief executive of the company in 1974, after two years as joint managing director.

**David Colville Foden** Eldest son of Ted Foden, born 1935. He was made an executive director in 1969, full director in 1970 and joint managing director in 1972. Specialist in marketing and export sales, like his father, and also involved in industrial matters.

**Stephen Patrick Twemlow** An engineer like his father, Edwin, born in 1938. A major contributor to Foden truck engineering from the mid-60s. Appointed an executive director in 1969, full director in 1970 and joint managing director in 1972.

**Edwin S. Foden** Second son of Ted Foden, born 1938, a great-grandson of the founder (like William and David). Appointed an executive director in 1971, specialising in sales, and particularly exports.

**Peter Foden** No connection with Fodens Ltd as a company, but as the youngest son of E. R. Foden he became an important family member in the ERF company in Sandbach. Born in 1930, he became managing director of ERF in 1960 when his older brother, Dennis, died.

**Top row, left to right**
*Edwin Foden, William Foden and Edwin Richard Foden.*

**Middle row, left to right**
*Samuel Poole Twemlow, James Edwin Foden, Reginald Gordon Foden and Edwin Twemlow.*

**Bottom row, left to right**
*William Louis Foden, David Colville Foden, Stephen Patrick Twemlow and Edwin S. Foden.*

# Index

*Page numbers printed in bold type refer to illustrations*

**A**
Ackerman steering 67
AEC 54, 139
    Mammoth 147
Ailsa 178
Albion Steel Co 26, 77, 121
Allchin, William 10, 13, 123
Alley & McLellan 78
Atkinson 123, 148, 159, 160, 161
Attlee, Clement 148
Austin 109, 137
    Cherub 120
    K2 137
    K3 137

**B**
*Bandmaster* 105, **106**
Bauly's 62
Bayley 40, 42
Beardmore power unit 120
Bedford 137, 162
    KD 58
    QL 137
Belle Vue Challenge Trophy 104, **104**
Bentley 77, 81
Bessemer 9
Besses o' th' Barn Band 101
Bignell, Vic 77
Black Dyke Mills Band 101, 108
Boer War 40, 99
Bourne, Albert 98
Bowen, J. 112
Bowyer, O. G. **142**, 146
Boyes, W. J. 112
brake technology 37, **49**, 65, 141, 144, **170**
Brakspear, W. H., & Sons 58

Brassington, Clifford 96
bridge restrictions 12, 15
brine-pumping engines 29
Bristol Company 149
Britain, E. W. M. 141
*Britannia* 76
*Britannia*, royal yacht 149
British Broadcasting Corporation 106
British National Brass Band Championship 101, 102, 103, 105, 106, 107, 108
British Open Brass Band Championship 101, 102, 103, 104, 108
British Road Services 149
Brown & May 42
Brunel 9, 10, 28
Brunner, Cecil H. 47, 50, 55, 87
Brunner Mond Ltd 47
Burrell, Charles, & Sons 10, 37, 40
buses **129**, 144-45, **144, 145, 146, 176,** 177-78, **178**
    FP (series 4) **143,** 144
    series 5 144

**C**
cab development 57, 65, 131, 137, **140, 142,** 152, 178
    S.18 **140,** 141, 146, **147**
    S.19 **140,** 146
    S.20 147
    S.21 (Sputnik) 153, **153**
    S.34 155, **156**
    S.36 **157,** 158
    S.39 155, 158, **172,** 173
    S.40 158, **158**
    S.50 **158,** 159

    S.60 159, **159**
    S.80 160, **177**
    S.80/83 **170,** 171, 173
    S.90 **166,** 171
    S.93 171, **176**
Calverley Mill 28
Centaur tank 136, **139**
chaff cutters 11
Chivers, Jacob 26, 28
Chivers, Thomas 27
Churchill, Winston 115, 116, 148
City of London 170, 171
Clarke & Co 29
Clarkson 40
Clayton 10, 13
Clayton & Shuttleworth 123
Clayton Tinplate Co 30
colliery winding-engines 23, 25, 26, 30
Colmore Refractories 156
Comet tank 136
Commercial Vehicle Show
    (1964) 155, 156
    (1966) 158
compound engine 14 *seq*; **16, 18, 19, 50**
computerisation 156, 162, 166
Construction and Use regulations 60, 155, 161, 166
Cooke, Tommy 75, **75**
corn-mill engine 30, 31
Coulthard 40
Cowapp, James 117
crane chassis **152, 153**
Creek St Engineering Co 42
Cresswell Colliery Band 101
Crossley 117
Crusader tank 135, 136, **139**
Cummins engines 152, 157, 158, 159, 160, 175, 176

**D**
Daf 166
*Daily Chronicle* 43, 45
*Daily Express* 45
*Daily Telegraph* 21, 45, 116
Daimler Benz 117, 162
    bus gearbox 120
Danks, Frederick 10
David Brown Gears 112
Delage 141
Dennis 64, 121, 131, 178
Department of Industry 168, 169, 170
diesel development 98, 120-25, 130-32, 133, 141-44
diesel engines and diesel engined chassis
    DG series **120, 130,** 131, **131,** 132, 133, 134, **136, 137, 138,** 141, 144, 146, 152
    FC series 152
    FC 35 157
    FD6 Mk III 153
    FD6 'Dynamic' 157
    FE series 146
    FE4 **141,** 143, 144
    FE4/8 **142,** 143
    FE6 143, **144**
    FG series 146, **147,** 154
    FGD dumpers 147
    FGT tractive units 149
    Fleetmaster **172,** 175, **176, 176**
    Haulmaster **172,** 176, **177**
    OG4 type 133, **134**
    R series 125, 127, 130, 133, 137
    S series 124, **128, 130, 132,** 133, 137, 149
    Sixer **172,** 173
    Super Haulmaster **176,** 177

2-tonner **121,** 130, 133
Universal **167,** 167, 174, 175
Dorman engines 109, 117, 121, 130
4DS 120
4JUR 120
'Drive into Europe' 76, **76,** 77, **77**
Dumpers **132,** 147, 149, **150, 151,** 152, 157, 158, 160, 176
Duramin 146

**E**
Eaton-Fuller 171, 173
Ebbw Vale Iron and Coal Co 25, 30
*Echo, The* 45
Elworth—*passim*
  George St 89, **91**
  Hill St 89
Elworth Foundry—*passim*
Elworth Silver Band 100, **100,** 101
'Emancipation Act' — see Locomotives on Highways Act (1896)
*Enterprise* 10
*Era* 11
ERF Ltd 96, 105, **111,** 112-14, 129, 131, 139, 147, 159, 161, 172

**F**
Fairey Aviation Band 108
Faulkner, George 111, 112
Faun Werk 152, 167
Festival of Britain 108
Fiat 166
fixed engines 12, 18, 22, 28, 30, 36
Foden & Hancock 13
Foden, David 170, **174,** 179, 180
Foden, Dennis 105, 109, 111, 112, 114, 125, 179
Foden, Edwin—*passim*; **6, 88, 100, 180**
Foden, Edwin (II) 179, **180**
Foden, E. R. 47 *seq;* **51, 100, 104, 113, 180**
Foden GmbH 177
Foden, Madge 112
Foden Motor Works Band 35, 60, 90, 99-108, **104, 107,** 139
Foden, Peter 179
Foden, Reg 57, 155, 157, 179, **180**

Foden, Ted 57, 126, 130, **154,** 155, 157, 179, **180**
Foden, William 13 *seq;* **71, 100, 104, 107, 154, 180**
Foden, William (Bill) **154,** 170, 172, 173, 179, **180**
Ford 162
Fowler, John 10, 13, 15, 40
Fuller Roadranger gearbox 175

**G**
Gardner engines 112, 117, 121, 131, 132, 144, 145, 148, 159, 160, 172, 176
  L2 98, 109, 120, 121, 133
  4LK 133
  LW 121, 133, 146
  4LW 105, 147
  5LW 137, **144**
  6LW 137, 147, 155
  6LX 155
  6LXB 155, 173
Garrett, Richard 10, 13, 40
General Strike 55, 94, 116, 117
Gibb, Sir George 62
Gibson, Ted 141
Gilbert, Fred 114
Girling brakes 173
Goddard, Freddy 76
Golland, John 108
'Great Polar Railway' 21
Guy 148
Gwyn, Charles 26

**H**
Hancock, George 9, 10, 11, 12, 13, 14, 22, 23, 33, 39, 46
Hancock, Walter 9, **10, 11**
Hanbury, John 26
Hardwick, Jack 76, 77, **77**
Harrison Saw Mills 26
Harvey's sherry 156
Hawker-Siddeley 172
Heath, Edward 77, 108
Heavy Motor Car Order of 1904 58, **59,** 60, 61
Hillhead Quarries 76
Hill Top Colliery 23, 34
Historic Commercial Vehicle Club 76
Home Guard (Fodens' Fusiliers) 140
House of Lords 61
housing 35, 36, 51, 56, 88, 89, **90, 91, 91,** 98, 119
Hughes Limestone Ltd 149
Hulme Colliery 25
Hurricane 134
Hynes, Thomas 103, 104

**I**
ICI 47
*Infant* 9, **11**
International Harvester 159
Invicta 77

**J**
jacknife 37
Jackson, Albert 117
Jackson, Alfred 101
Jacksons, hauliers 121
Jarrow March 96
Jennings, J. H. 112
Johnson, Ted 130, **154**
Johnstone, Bob 141
Jones, William, Jute Sack Co 26, 33
Judkins, Ken 73, 74, 75
jute-mill engine 26

**K**
Kidwelly Tinplate Co 26, **27, 28, 29,** 33
King Edward VII 100
King George V 57, 102
King George VI 106
Kirkstall Forge Co 112, 146

**L**
labour relations 33-36, 88-98
Lea Bros 30, **31**
Leyland 40, 52, 54, 63, 64, 70, 76, 108, 121, 131, 136, 139, 145, 161, 178
  Atlantean 144
  Beaver 147
  Cub 133
  Hippo 147
  Lynx 133
  Octopus 147, 149
  Terrier 58
Liberty engine 135
Lipe Rollway clutch 175
List-Brain, Mike 76
*Little Giant* 37
Liverpool Self-propelled Traffic Association 39
Liverpool Trials 39, 40
Lloyd George 62, 63
Locomotives Acts 12
  (1861) 12
  (1865) 12
Locomotive and Threshing Engines Act (1894) 21
Locomotives on Highways Act (1896) 21, 33, 39, 58

**M**
macadam 61
MacDonald, Ramsey 122

Magirus 166
Mam Tor 67, 86
MAN 117, 166
*Manchester Mercury* 32, 33
Mann's Patent Steam Cart and Waggon Co 39, 40, 123
'Mann Steam Cart' 40
Marshall 40
Mason, Edward 119
Mason, Ernest 36
Mason, Fred 23, 25, 33, **36,** 88, 101, 141
Mason, George 101
Mason, Harry 141
Maudslay Mogul 48
Mays, Raymond 141
McAdam, John 12
Meadows 4EL engine 120
Mercedes 166
Metcalf, John 12
Metrobus 178
MHS International Ltd 166
Mills, Jack 141
Milnes, George F., & Co 42, 43
Ministry of Transport 136
Ministry of War Production 135
Morris Motors 108, 137
  bullnose 77
  tonner 76
Mortimer, Alex 104
Mortimer, Fred 103, 104, 105, 107, **107,** 108
Mortimer, Harry 104, 105, 107, **107,** 108
Mortimer, Rex 104, **107,** 108
Motor Panels Ltd 158, 171
Motor Show (1933) 112
  (1935) 129
Mow Cop 67
munitions production **51,** 53, 90, 103, 106, 134-35, 139

**N**
Naylers 10
Norris, William 61
Nuffield group 135
Nuffield Mechanisation Aero Group 135

**P**
Palmer Mann Salt Co 81
Parmiter, P. J. 39, 40
Parsons 9
Pickfords Ltd 53, 62
Pierpoint 141
Place, Francis 32
Plant & Hancock 9
Poole, Alcock & Co 75

*Index*

Poole, Francis 51, 55, 56, 109, 126, 128
portable engines 11, 13, 14, 15, 16, 22, 49
Price Waterhouse 169, 170
*Pride of Edwin* 155
*Pride of Leven* **18**
*Prospector* 19, **20**
*Puffing Billy* 76, 102, 103, **103**, 105
*Punch* 21

**Q**

Queen Elizabeth II 106
Queen Victoria 9, 12, 37, 50

**R**

Radcliffe, Goff 77
railways, working conditions 32, 33
Ransomes 10, 13, 40
red flag 12, 15, 17, 21, 37
Renolds chain 64, 67
Rimmer, William 101, 102
Rivett, Bill 76, 77
Road and Rail Traffic Act (1933) 123, 137
Road Board Scheme 62
Road Fund 115, 116
Road Traffic Act (1930) 122
Robey 123
Rockwell 171, 175
Rolls Royce 158, 160, 172, 175
   Eagle diesel 157
Royal Agricultural Society
   Show (1883) 17
   Trials (1887) **18**
Rubery Owen 146
Russell & Newbury 120
Ruston 135

**S**

Salter Report 110, 123, 124
Salter, Sir Arthur 122
Sandbach — *passim*
*Sandbach Chronicle* 168
Sankey Co 30
Saviem 166
Scania-Vabis 162, 163, 166
Scottish Motor Show 105 (1971) 161
Seddon 133, 148, 159, 161
Sentinel 64, 78, 83, 86, 109, 123, 149, 157
Sherratt, Ernest 110, 111, 112, 119
Smallwood 11, 33, 102
social club 35, 88, 91
speed limits 12, 15, 21, 37, 58, 59, 60, 87, 147
Spitfire 134
steam wagon development 37 *seq;* **38, 39, 40, 41, 48**
   'Agitractor' 84, **123**
   C-type **54, 55,** 56, **57,** 63-67, **63, 65,** 68, 73, 76, **76,** 78, 83, 137
   driving and maintenance 69-77, 96
   D-type **53, 66,** 68, **68,** 74, 76
   E-type **79,** 80, **80,** 81, 83, 115, **118, 122**
   5-tonners **51,** 52, **52,** 54, 57, 60, 62, **71, 72,** 73
   flexible six wheeler 68
   N-type 84, 85, **116, 117**
   overtype 39, 46, 57, 64, 68, 73, 78, 80, 84, 96, 109, 116
   Q-type 84, 85, **86,** 119
   Speed-Six (O-type) **74,** 80, 81, **81, 84, 85,** 109, 116, 120, 137
   Speed-Twelve **74, 82,** 83, **83, 84,** 85, 116, 120
   'Sun' tractor 84, 85, **122**
   3-tonners **49,** 73, 74, 75, **100**
   12-tonners 67, 68, 73, 76
   undertype 40, **43,** 46, 57, 68, 73, 78, 83, 84, 96, 109, 115, 116, 120
Stephenson, George 9, 10, 28
Stephenson-Howe 64
Stockton, Sir Edwin 127, 132
Stour Valley Co 24
Straker Steam Vehicle Co 42, 43, 44, 45
straw baler 19
Stubbs Brothers 28, 47
Stubbs, John 49, 50, 117, 127

**T**

tar macadam 61, 62
Tasker, William 9, 13, 37, 40
Telford, Thomas 62
Thornycroft 40, 42, 43, **43,** 44, **44,** 45, 52, 54, 70, 121, 131, 149
threshing machines 11, **14,** 16, 17, 24
tilling machinery 11
*Times, The* 21, 45
Tin plate production 26-28
   engines 36
Tolley, Leslie J. 171
Transport Act (1947) 148
*Truck* 172, 173
Twemlow, Eddie 141, **154,** 155, 179, **180**
Twemlow, Patrick 179, **180**
Twemlow, S. P. 47, 49, 55, 101, 109, 115, 117, 179, 180

**V**

Vernon Tinplate Co 30
Veteran Car Club 76
Vintage Motor Cycle Club 76
Vintage Sports Car Club 76
Volvo 166
Vulcan 149

**W**

wages 33, 88, 91, 95, **95, 96,** 104, 109, 167
Wallis & Steevens 10, 37, 60, 61
Wantage Engineering Co 42
War Office 51, 53, 62, 90, 106, 135, 137
War Office Committee on Mechanical Transport 40, 41, 46
War Office Trials 34, 40-46, **41, 44, 45,** 58, 60, **100,** 102, 134
Webb compound locomotive 28
Wheelock 99
Whitecross Wire Co 23
Whittle Superheater Co 125, 126, 127, 128
Whittle, Wood 109, 121, 122, 125, 126
Whitworth 9
Williams, Charlie 76
Wilson, Harold 169
winch engine 30
wire-drawing mill engine 23, 30
working conditions 33, 36, 94, 95, 98, 103, 134, 135
Wormald, Edward 102

**Y**

Yorkshire 64, 70

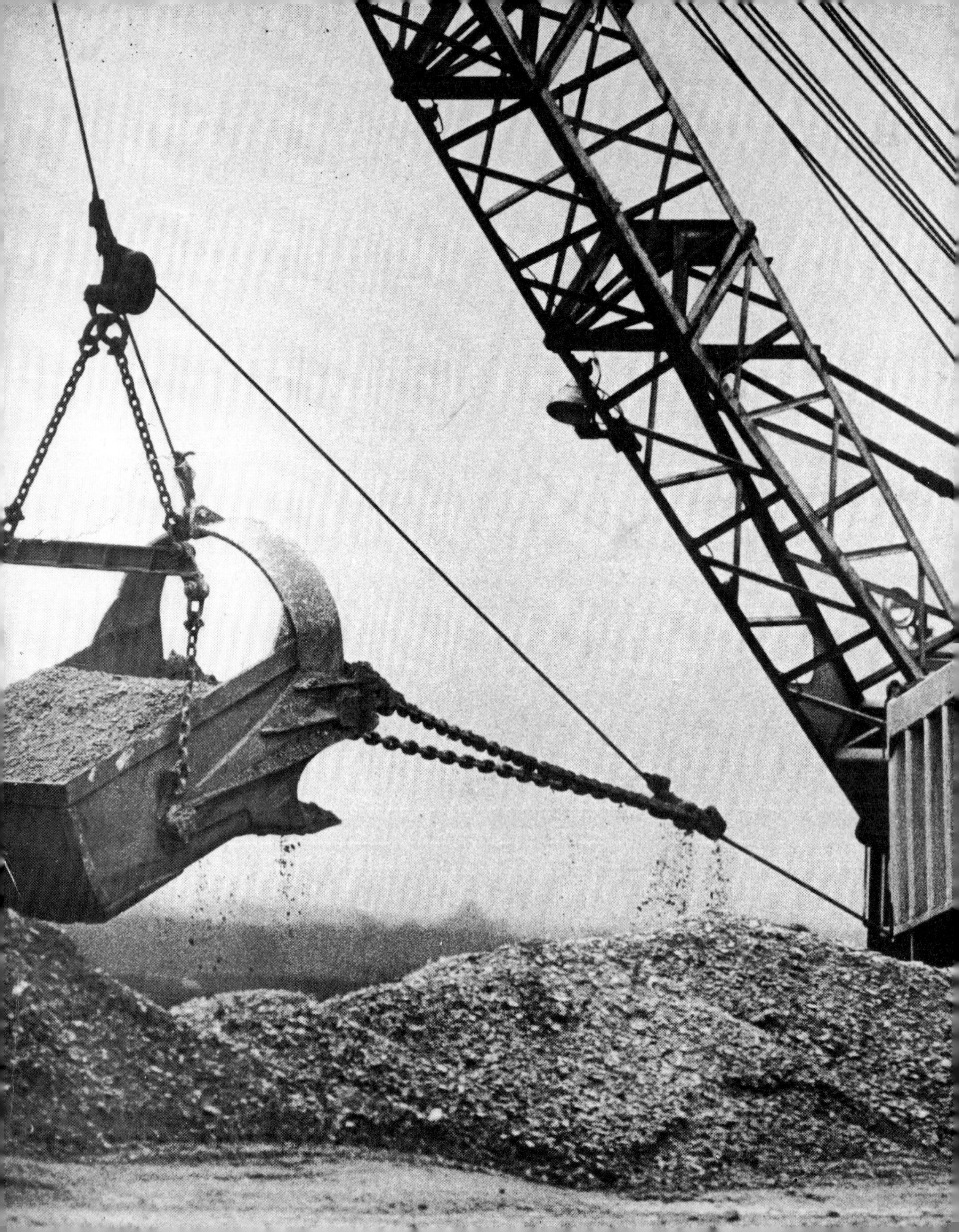